ISBN 978-0-266-86682-4
PIBN 10901687

UCID-
Vol. 2

Seismic Hazard Characterizati
of 69 Nuclear Plant Sites
East of the Rocky Mountains

Results and Discussion for the Batch 1 Sites

Prepared by D. L. Bernreuter, J. B. Savy, R. W. Mensing, J. C. Chen

Lawrence Livermore National Laboratory

Prepared for
U.S. Nuclear Regulatory
Commission

NOTICE

Availability of Reference Materials Cited in NRC Publications

Most documents cited in NRC publications will be available from one of the following sources:

1. The NRC Public Document Room, 1717 H Street, N.W.
   Washington, DC 20555

2. The Superintendent of Documents, U.S. Government Printing Office, Post Office Box 37082, Washington, DC 20013-7082

3. The National Technical Information Service, Springfield, VA 22161

Although the listing that follows represents the majority of documents cited in NRC publications, it is not intended to be exhaustive.

Referenced documents available for inspection and copying for a fee from the NRC Public Document Room include NRC correspondence and internal NRC memoranda; NRC Office of Inspection and Enforcement bulletins, circulars, information notices, inspection and investigation notices; Licensee Event Reports; vendor reports and correspondence; Commission papers; and applicant and licensee documents and correspondence.

The following documents in the NUREG series are available for purchase from the GPO Sales Program: formal NRC staff and contractor reports, NRC-sponsored conference proceedings, and NRC booklets and brochures. Also available are Regulatory Guides, NRC regulations in the *Code of Federal Regulations*, and *Nuclear Regulatory Commission Issuances.*

Documents available from the National Technical Information Service include NUREG series reports and technical reports prepared by other federal agencies and reports prepared by the Atomic Energy Commission, forerunner agency to the Nuclear Regulatory Commission.

Documents available from public and special technical libraries include all open literature items, such as books, journal and periodical articles, and transactions. *Federal Register* notices, federal and state legislation, and congressional reports can usually be obtained from these libraries.

Documents such as theses, dissertations, foreign reports and translations, and non-NRC conference proceedings are available for purchase from the organization sponsoring the publication cited.

Single copies of NRC draft reports are available free, to the extent of supply, upon written request to the Division of Information Support Services, Distribution Section, U.S. Nuclear Regulatory Commission, Washington, DC 20555.

Copies of industry codes and standards used in a substantive manner in the NRC regulatory process are maintained at the NRC Library, 7920 Norfolk Avenue, Bethesda, Maryland, and are available there for reference use by the public. Codes and standards are usually copyrighted and may be purchased from the originating organization or, if they are American National Standards, from the American National Standards Institute, 1430 Broadway, New York, NY 10018.

Seismic Hazard Characterization of 69 Nuclear Plant Sites East of the Rocky Mountains

Results and Discussion for the Batch 1 Sites

Manuscript Completed: November 1988
Date Published: January 1989

Prepared by
D. L. Bernreuter, J. B. Savy, R. W. Mensing, J. C. Chen

Lawrence Livermore National Laboratory
7000 East Avenue
Livermore, CA 94550

Prepared for
Division of Engineering and System Technology
Office of Nuclear Reactor Regulation
U.S. Nuclear Regulatory Commission
Washington, DC 20555
NRC FIN A0448

Seismic Hazar[illegible] of 69 Nuclear Plan[illegible] East of the Rocky [illegible]

Results and Discu[illegible]

Abstract

The EUS Seismic Hazard Characterization Project (SHC) is the outgrowth of an earlier study performed as part of the U.S. Nuclear Regulatory Commission's (NRC) Systematic Evaluation Program (SEP). The objectives of the SHC were: (1) to develop a seismic hazard characterization methodology for the region east of the Rocky Mountains (EUS), and (2) the application of the methodology to 69 site locations, some of them with several local soil conditions. The method developed uses expert opinions to obtain the input to the analyses. An important aspect of the elicitation of the expert opinion process was the holding of two feedback meetings with all the experts in order to finalize the methodology and the input data bases. The hazard estimates are reported in terms of peak ground acceleration (PGA) and 5% damping velocity response spectra (PSV).

A total of eight volumes make up this report which contains a thorough description of the methodology, the expert opinion's elicitation process, the input data base as well as a discussion, comparison and summary volume (Volume VI).

Consistent with previous analyses, this study finds that there are large uncertainties associated with the estimates of seismic hazard in the EUS, and it identifies the ground motion modeling as the prime contributor to those uncertainties.

The data bases and software are made available to the NRC and to public uses through the National Energy Software Center (Argonne, Illinois).

Volume II

List of Tables and Figures

The same format for the tables and figures are used for every site. The following is an exhaustive list of all tables and figures presented in this volume.

The symbol "SN" in the following refers to "Site Number" and the corresponding page numbers are given in table page xii.

List of Additional Tables and Figures

Site
SN

Site												
1. Fitzpatrick	14	15	16	17	18	19	20	21	22	23	24	25
2. Ginna	28	29	30	31	32	33	34	35	36	37	38	
3. Haddam Neck	40	41	42	43	44	45	46	47	48	49	50	
4. Hope Creek	52	53	54	55	56	57	58	59	60	61	62	63
5. Indian Point	65	66	67	68	69	70	71	72	73	74	75	
6. Limerick	77	78	79	80	81	82	83	84	85	86	87	
7. Maine Yankee	89	90	91	92	93	94	95	96	97	98	99	
8. Millstone	101	102	103	104	105	106	107	108	109	110	111	
9. Nine Mile Pt	113	114	115	116	117	118	119	120	121	122	123	
10. Oyster Creek	125	126	127	128	129	130	131	132	133	134	135	
11. Peach Bottom	137	138	139	140	141	142	143	144	145	146	147	
12. Pilgrim	149	150	151	152	153	154	155	156	157	158	159	
13. Salem	161	162	163	164	165	166	167	168	169	170	171	
14. Seabrook	173	174	175	176	177	178	179	180	181	182	183	
15. Shoreham	185	186	187	188	189	190	191	192	193	194	195	
16. Susquehanna	197	198	199	200	201	202	203	204	205	206	207	208
17. Three Mile Island	210	211	212	213	214	215	216	217	218	219	220	
18. Vermont Yankee	222	223	224	225	226	227	228	229	230	231	232	233
19. Yankee at Rowe	235	236	237	238	239	240	241	242	243	244	245	246

Foreword

The impetus for this study came from two unrelated needs of the Nuclear Regulatory Commission (NRC). One stimulus arose from the NRC funded "Seismic Safety Margins Research Programs" (SSMRP). The SSMRP's task of simplified methods needed to have available data and analysis software necessary to compute the seismic hazard at any site located east of the Rocky Mountains which we refer to as the Eastern United States (EUS) in a form suitable for use in probabilistic risk assessment (PRA). The second stimulus was the result of the NRC's discussions with the U.S. Geological Survey (USGS) regarding the USGS's proposed clarification of their past position with respect to the 1886 Charleston earthquake. The USGS clarification was finally issued on November 18, 1982, in a letter to the NRC, which states that:

> "Because the geologic and tectonic features of the Charleston region are similar to those in other regions of the eastern seaboard, we conclude that although there is no recent or historical evidence that other regions have experienced strong earthquakes, the historical record is not, of itself, sufficient ground for ruling out the occurrence in these other regions of strong seismic ground motions similar to those experienced near Charleston in 1886. Although the probability of strong ground motion due to an earthquake in any given year at a particular location in the eastern seaboard may be very low, deterministic and probabilistic evaluations of the seismic hazard should be made for individual sites in the eastern seaboard to establish the seismic engineering parameters for critical facilities."

Anticipation of this letter led the Office of Nuclear Reactor Regulation to jointly fund a project with the Office of Nuclear Regulatory Research. The results were presented in Bernreuter et. al., (1985), and the objectives were:

1. to develop a seismic hazard characterization methodology for the entire region of the United States east of the Rocky Mountains.

2. to apply the methodology to selected sites to assist the NRC staff in their assessment of the implications in the clarification of the USGS position on the Charleston earthquake, and the implications of the occurrence of the recent earthquakes such as that which occurred in New Brunswick, Canada, in 1982.

The methodology used in that 1985 study evolved from two earlier studies that the Lawrence Livermore National Laboratory (LLNL) performed for the NRC. One study, Bernreuter and Minichino (1983), was part of the NRC's Systematic Evaluation Program (SEP) and is simply referred hereafter to as the SEP study. The other study was part of the SSMRP.

At the time (1980-1985), an improved hazard analysis methodology and EUS seismicity and ground motion data set were required for several reasons:

- o Although the entire EUS was considered at the time of the SEP study, attention was focused on the areas around the SEP sites--mainly in the Central United States (CUS) and New England. The zonation of other areas was not performed with the same level of detail.

- o The peer review process, both by our Peer Review Panel and other reviewers, identified some areas of possible improvements in the SEP methodology.

- o Since the SEP zonations were provided by our EUS Seismicity Panel in early 1979, a number of important studies had been completed and several significant EUS earthquakes had occurred which could impact the Panel members' understanding of the seismotectonics of the EUS.

- o Our understanding of the EUS ground motion had improved since the time the SEP study was performed.

By the time our methodology was firmed up, the expert opinions collected and the calculations performed (i.e. by 1985), the Electric Power Research Institute (EPRI) had embarked on a parallel study.

We performed a comparative study, Bernreuter et. al., (1987), to help in understanding the reasons for differences in results between the LLNL and the EPRI studies. The three main differences were found to be: (1) the minimum magnitude value of the earthquakes contributing to the hazard in the EUS, (2) the ground motion attenuation models, and (3) the fact that LLNL accounted for local site characteristics and EPRI did not. Several years passed between the 1985 study and the application of the methodology to all the sites in the EUS. In recognition of the fact that during that time a considerable amount of research in seismotectonics and in the field of strong ground motion prediction, in particular with the development of the so called random vibration or stochastic approach, NRC decided to follow our recommendations and have a final round of feedback with all our experts prior to finalizing the input to the analysis.

In addition, we critically reviewed our methodology which lead to minor improvements and we also provided an extensive account of documentation on the ways the experts interpreted our questionnaires and how they developed their answers. Some of the improvements were necessitated by the recognition of the fact that the results of our study will be used, together with results from other studies such as the EPRI study or the USGS study, to evaluate the relative hazard between the different plant sites in the EUS.

This report includes eight volumes:

Volume I provides an overview of the methodology we developed for this project. It also documents the final makeup of both our Seismicity and

Ground Motion Panels, and documents the final input from the members of both panels used in the analysis. Comparisons are made between the new results and previous results.

Volumes II to V provide the results for all the active nuclear power plant sites of the EUS divided into four batches of approximately equal size and of sites roughly located in the four main geographical regions of the EUS (NE, SE, NC and SC). A regional discussion is given in each of Vols. II to V.

Volume VI emphasizes important sensitivity studies, in particular the sensitivity of the results to correction for local site conditions and G-Expert 5's ground motion model. It also contains a summary of the results and provides comparisons between the sites within a common region and for sites between regions.

Volume VII contains unaltered copies of the ten questionnaires used from the beginning of the 1985 study to develop the complete input for this analysis.

After the bulk of the work was completed and draft reports for Vols. I-VII were written, additional funding became available.

Volume VIII contains the hazard result for the 12 sites which were primarily rock sites but which also had some structures founded on shallow soil. These results supplement the results given in Vols. II to V where only the primary soil condition at the site was used.

List of Abbreviations and Symbols

A	Symbol for Seismicity Expert 10 in the figures displaying the results for the S-Experts
ALEAS	Computer code to compute the BE Hazard and the CP Hazard for each seismicity expert
AM	Arithmetic mean
AMHC	Arithmetic mean hazard curve
B	Symbol for Seismicity Expert 11 in the figures displaying the results for the S-Experts
BE	Best estimate
BEHC	Best estimate hazard curve
BEUHS	Best estimate uniform hazard spectrum
BEM	Best estimate map
C	Symbol for Seismicity Expert 12 in the figures displaying the results for the S-Experts
COMAP	Computer code to generate the set of all alternative maps and the discrete probability density of maps
COMB	Computer code to combine BE hazard and CP hazard over all seismicity experts
CP	Constant percentile
CPHC	Constant percentile hazard curve
CPUHS	Constant percentile uniform hazard spectrum
CUS	Central United States, roughly the area bounded in the west by the Rocky Mountains and on the east by the Appalachian Mountains, excluding both mountain systems themselves
CZ	Complementary zone
D	Symbol for Seismicity Expert 13 in the figures displaying the results for the S-Experts
EPRI	Electric Power Research Institute

EUS	Used to denote the general geographical region east of the Rocky Mountains, including the specific region of the Central United States (CUS)
g	Measure of acceleration: 1g = 9.81m/s/s = acceleration of gravity
G-Expert	One of the five experts elicited to select the ground motion models used in the analysis
GM	Ground motion
HC	Hazard curve
I_o	Epicentral intensity of an earthquake relative to the MMI scale
I_s	Site intensity of an earthquake relative to the MMI scale
LB	Lower bound
LLNL	Lawrence Livermore National Laboratory
M	Used generically for any of the many magnitude scales but generally m_b, m_b(Lg), or M_L.
M_L	Local magnitude (Richter magnitude scale)
M_b	True body wave magnitude scale, assumed to be equivalent to m_b(Lg) (see Chung and Bernreuter, 1981)
m_b(Lg)	Nuttli's magnitude scale for the Central United States based on the Lg surface waves
M_S	Surface wave magnitude
MMI	Modified Mercalli Intensity
M_o	Lower magnitude of integration. Earthquakes with magnitude lower than M_o are not considered to be contributing to the seismic hazard
NC	North Central; Region 3
NE	North East; Region 1
NRC	Nuclear Regulatory Commission
PGA	Peak ground acceleration
PGV	Peak ground velocity

PRD Computer code to compute the probability distribution of epicentral distances to the site

PSRV Pseudo relative velocity spectrum. Also see definition of spectra below

Q Seismic quality factor, which is inversely proportional to the inelastic damping factor.

Q1 Questionnaire 1 - Zonation (I)

Q2 Questionnaire 2 - Seismicity (I)

Q3 Questionnaire 3 - Regional Self Weights (I)

Q4 Questionnaire 4 - Ground Motion Models (I)

Q5 Questionnaire 5 - Feedback on seismicity and zonation (II)

Q6 Questionnaire 6 - Feedback on ground motion models (II)

Q7 Questionnaire 7 - Feedback on zonation (III)

Q8 Questionnaire 8 - Seismicity input documentation

Q9 Questionnaire 9 - Feedback on seismicity (III)

Q10 Questionnaire 10 - Feedback on ground motion models (III)

R Distance metric, generally either the epicentral distance from a recording site to the earthquake or the closest distance between the recording site and the ruptured fault for a particular earthquake.

Region 1 (NE): North East of the United States, includes New England and Eastern Canada

Region 2 (SE): South East United States

Region 3 (NC): North Central United States, includes the Northern Central portions of the United States and Central Canada

Region 4 (SC): Central United States, the Southern Central portions of the United States including Texas and Louisiana

RP Return period, in years

RV Random vibration. Abbreviation used for a class of ground motion models also called stochastic models.

S	Site factor used in the regression analysis for G-Expert 5's GM model: S = 0 for deep soil, S = 1 for rock sites
SC	South Central; Region 4
SE	South East; Region 2
S-Expert	One of the eleven experts who provide the zonations and seismicity models used in the analysis
SEP	Systematic Evaluation Program
SHC	Seismic Hazard Characterization
SHCUS	Seismic Hazard Characterization of the United States
SN	Site Number
Spectra	Specifically in this report: attenuation models for spectral ordinates were for 5% damping for the pseudo-relative velocity spectra in PSRV at five frequencies (25, 10, 5, 2.5, 1 Hz).
SSE	Safe Shutdown Earthquake
SSI	Soil-structure-interaction
SSMRP	Seismic Safety Margins Research Program
UB	Upper bound
UHS	Uniform hazard spectrum (or spectra)
USGS	United States Geological Survey
WUS	The regions in the Western United States where we have strong ground motion data recorded and analyzed
γ	Regional absorption coefficient

Executive Summary: Volume II

The impetus for this study came from two unrelated needs of the Nuclear Regulatory Commission (NRC). One stimulus arose from the NRC funded "Seismic Safety Margins Research Programs" (SSMRP). The SSMRP's task of simplified methods needed to have available data and analysis software necessary to compute the seismic hazard at any site located in the eastern United States (EUS) in a form suitable for use in probabilistic risk assessment (PRA). The second stimulus was the result of the NRC's discussions with the U.S. Geological Survey (USGS) regarding the USGS's proposed clarification of their past position with respect to the 1886 Charleston earthquake. The USGS clarification was finally issued on November 18, 1982, in a letter to the NRC, which states that:

> "Because the geologic and tectonic features of the Charleston region are similar to those in other regions of the eastern seaboard, we conclude that although there is no recent or historical evidence that other regions have experienced strong earthquakes, the historical record is not, of itself, sufficient ground for ruling out the occurrence in these other regions of strong seismic ground motions similar to those experienced near Charleston in 1886. Although the probability of strong ground motion due to an earthquake in any given year at a particular location in the eastern seaboard may be very low, deterministic and probabilistic evaluations of the seismic hazard should be made for individual sites in the eastern seaboard to establish the seismic engineering parameters for critical facilities."

Anticipation of this letter led the Office of Nuclear Reactor Regulation to jointly fund a project with the Office of Nuclear Regulatory Research. The results were presented in Bernreuter et al. in 1985 and the objectives were:

1. to develop a seismic hazard characterization methodology for the entire region of the United States east of the Rocky Mountains (Referred to as EUS in this report).

2. to apply the methodology to selected sites to assist the NRC staff in their assessment of the implications in the clarification of the USGS position on the Charleston earthquake, and the implications of the occurrence of the recent eastern U.S. earthquakes in New Brunswick and New Hampshire.

The methodology used in that 1985 study evolved from two earlier studies LLNL performed for the NRC. One study, Bernreuter and Minichino (1983), was part of the NRC's Systematic Evaluation Program (SEP) and is simply referred hereafter to as the SEP study. The other study was part of the SSMRP.

At the time (1980-1985), an improved hazard analysis methodology and EUS seismicity and ground motion data set were required for several reasons:

- Although the entire EUS was considered at the time of the SEP study, attention was focused on the areas around the SEP sites--mainly in the Central United States (CUS) and New England. The zonation of other areas was not performed with the same level of detail.

- The peer review process, both by our Peer Review Panel and other reviewers, identified some areas of possible improvements in the SEP methodology.

- Since the SEP zonations were provided by our EUS Seismicity Panel in early 1979, a number of important studies have been completed and several significant EUS earthquakes have occurred which could impact the Panel members' understanding of the seismotectonics of the EUS.

- Our understanding of the EUS ground motion had improved since the time the SEP study was performed.

By the time our methodology was firmed up, the expert opinions collected and the calculations performed (i.e. by 1985), the Electric Power Research Institute (EPRI) had embarked in a paralleled study.

We performed a comparative study (Bernreuter et al. 1987) whose purpose was to help in understanding the reasons for differences in results between the LLNL and the EPRI study (EPRI 1985a and 1985b). The three main differences were found to be (1) the minimum magnitude value of the earthquakes contributing to the hazard in the EUS, (2) the ground motion attenuation models, and (3) the fact that LLNL accounted for local site characteristics and EPRI did not. Several years passed between the 1985 study and the time when NRC actually decided to apply the methodology to all the sites in the EUS. In recognition of the fact that during that time a considerable amount of research in seismotectonics and in the field of strong ground motion prediction, in particular with the development of the so called random vibration or stochastic approach, NRC decided to follow our recommendations and have a final round of feedback with all our experts prior to finalizing the input to the analysis.

In addition, we critically reviewed our methodology which lead to minor improvements and we also provided an extensive account of documentation on the ways the experts interpreted our questionnaires and how they developed their answers. Some of the improvements were necessitated by the recognition of the fact that the results of our study will be used, together with results from other studies such as the EPRI study or the USGS study, to evaluate the relative hazard between the different plant sites in the EUS.

This volume (Volume II) is one of eight volumes where the methodology and the results of the analysis are presented. The analysis was performed for a total of 69 different geographic locations. These sites were divided into four groups (batches) of approximately equal size. Volume III presents the results for the group of sites roughly located in the Southeastern part of the EUS. It contains 17 sites, Vol. IV concerns sites located in the North Central EUS and Vol. V concerns the remaining sites. In addition, Vol. VIII represents the results for an alternative soil condition at some of the sites.

The results are presented individually for each site together with comments. The seismic hazard results presented here account for earthquakes of magnitude 5 or above only, and a set of calculations was made to provide an estimate of the seismic hazard created at each of the sites by the earthquakes of magnitude between 3.75 and 5.

In addition, a discussion on uncertainty, comparison between sites, sensitivity to site location and a discussion on the factors influencing the distribution of the contributing zones, is presented in Section 3.

The other volumes provide an extensive description of the methodology (Volume I), the results for the other groups of sites (Volume II, IV and V), a summary and some comparisons between sites and groups of sites (Volume VI), and finally a copy of all the questionnaires used in the analysis to develop the input is given in Volume VII.

1. INTRODUCTION

In this Volume we present the seismic hazard estimates for the 19 sites in Batch 1 listed in Table 1.1 and plotted in Fig. 1.1. The seismic hazard results for the Batch 1 sites are based on:

- The zonation and seismicity inputs provided by our 11 seismicity experts (S-Experts) listed in Table 1.2.
- The ground motion models (peak ground acceleration (PGA) models and 5 percent damped pseudo velocity spectral models) provided by our 5 G-Experts listed in Table 1.3.
- The methodology we developed and described in Vol. I of this report.

The results presented in his report differ from our previous results Bernreuter et al. (1984, 1985) for the following reasons:

- Used the final updated input from our S and G-Experts given in Section 3 and Appendix B of Vol. I. As discussed in Vol. I, S-Experts 3,6,7 and 12 provided completely new zonations and seismicity parameters, S-Experts 4,10,11 and 13 modified some of their zones and seismicity parameters, and the G-Experts significantly changed their ground motion models.
- Used $m_b = 5.0$ as the lower bound of integration, i.e., we only included the contribution from earthquakes with magnitude 5.0 and greater. In our previous reports, Bernreuter et al. (1984, 1985), we used a lower bound of 3.75. This change has a significant impact on the results as discussed in Bernreuter et al. (1985). We present plots for every site which gives an indication of how much the earthquakes in the magnitude range 3.75 to 5 would contribute to the seismic hazard estimates given in the report.

Corrections for the soil conditions at each site have been included using the approach outlined in Section 3.7 of Vol. I. However, it is important to note that each site is put in a single fixed site soil category as listed in Table 1.1. At some sites the main containment buildings are founded on rock but some structures are founded on shallow soil. The seismic hazard estimates given in this report are at the free surface and for rock sites it is at the free surface of the rock with the soil removed. Thus for these rock sites which have a few structures founded on shallow soil the results presented here should be corrected for the shallow soil amplification effects as described in Vol. VI before they are applied to the structures founded on soil. If all structures are founded on the same soil condition, then no added correction is needed.

Section 2 of this report contains the results for each site and, some site specific discussion. In section 3 of this report we make regional observations and comparisons between sites. In Vol. VI we reach overall conclusions based on the regional results presented in this volume and Vols. III-V.

TABLE 1.1

SITES AND SOIL CATEGORY USED FOR EACH SITE IN BATCH 1

	SITE NAME	MAP (1) KEY	SOIL CATEGORY (2)
1.	Fitzpatrick	1	Rock
2.	Ginna	2	Rock
3.	Haddam Neck	3	Rock
4.	Hope Creek	4	Deep Soil
5.	Indian Point	5	Rock
6.	Limerick	6	Rock
7.	Maine Yankee	7	Rock
8.	MillStone	8	Rock
9.	Nine Mile Pt.	9	Rock**
10.	Oyster Creek	A	Deep Soil
11.	Peach Bottom	B	Rock
12.	Pilgrim	C	Sand-Like 2
13.	Salem	D	Deep Soil
14.	Seabrook	E	Rock
15.	Shoreham	F	Deep Soil
16.	Susquehanna	G	Rock**
17.	Three Mile Island	H	Rock**
18.	Vermont Yankee	I	Rock
19.	Yankee at Rowe	J	Till- Like 2

(1) Key used on Fig. 1.1

(2) Site categories are given in Table 3.9 of Vol. I and repeated in Table 1.4.

(**) Have some structures founded on shallow soil.

TABLE 1.2

FINAL EUS ZONATION AND SEISMICITY PANEL MEMBERS
(S-Panel)

Professor Gilbert A. Bollinger

Mr. Richard J. Holt

Professor Arch C. Johnston

Dr. Alan L. Kafka

Professor James E. Lawson

Professor L. Tim Long

Professor Otto W. Nuttli

Dr. Paul W. Pomeroy

Dr. J. Carl Stepp

Professor Ronald L. Street

Professor M. Nafi Toksöz

TABLE 1.3

FINAL EUS GROUND MOTION MODELING PANEL MEMBERS
(G-Panel)

Dr. David M. Boore

Dr. Kenneth Campbell

Professor Mihailo Trifunac

Dr. John Anderson

Dr. John Dwyer

TABLE 1.4
DEFINITION OF THE EIGHT SITE CATEGORIES

		CATEGORY	DEPTH
Generic Rock			
(1)		Rock	N/A
Sand Like			
(2)	Sand 1	S1	25 to 80 ft.
(3)	Sand 2	S2	80 to 180 ft.
(4)	Sand 3	S3	180 to 300 ft.
Till-Like			
(5)	Till 1	T1	25 to 80 ft.
(6)	Till 2	T2	80 to 180 ft.
(7)	Till 3	T3	180 to 300 ft.
Deep Soil			
(8)		Deep Soil	N/A

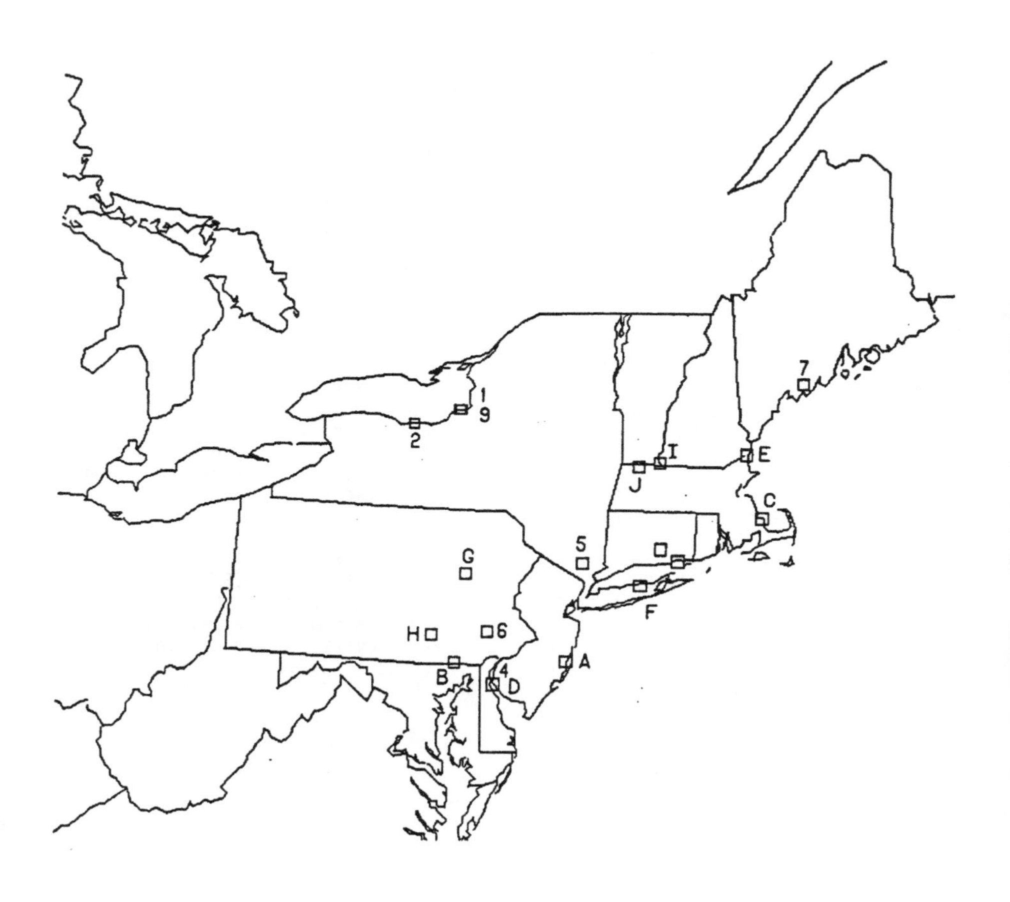

Figure 1.1 Map showing the location of the Batch 1 sites contained in Vol. II of this report. Map symbols are given in Table 1.1.

2. RESULTS AND SITE SPECIFIC DISCUSSION

2.0 General Introduction

In sections 2.1 to 2.19 we provide the results for the sites listed in Table 1.1. Using a uniform format for each site (i.e. each section), we first present a table providing the following information:

- o the soil category used in the analysis to correct for local site conditions.

- o For each S-Expert the table provides a listing of the four seismic zones which contribute most to the hazard in terms of the peak ground acceleration (PGA) hazard at both lower PGA (0.125g) and at higher PGA (0.6g) values. The zone ID's listed in the tables are keyed to the S-Experts' maps given in Appendix B of this report.

The contribution of various zones given in the table for each site is limited only to the contribution to the best estimate hazard curves (BEHCs). That is, only the zones on the best estimate map (i.e. those zones which have a probability of existence of 0.5 or greater) and only the best estimate (BE) ground motion attenuation models for the peak ground acceleration (PGA) are used. This is a limitation that should be kept in mind as in a few cases contributing zones with a probability of existence of less than 0.5 will not be listed. In addition, the contributing zones for the uniform hazard spectra might be different from the contribution for PGA, and also it might be different between frequencies.

The table, labeled Table 2.SN.1, is followed by ten figures, 2.SN.1 to 2.SN.10 (SN = Site Number given in Table 1.1). The first four figures, Figs. 2.SN.1 - 2.SN.4 give various PGA hazard curves. The next six figures, Figs. 2.SN.5 - 2.SN.10 give various 5 percent damped relative velocity spectra for various return periods. It should be noted that the spectra result from straight lines drawn between the five periods, (0.04s, 0.1s, 0.2s, 0.4s and 1.0s) at which the calculations were performed.

Figures 2.SN.1 give a comparison between the best estimate hazard curve (BEHC), and the arithmetical mean hazard curve (AMHC) for the peak ground acceleration (PGA).
The BEHC and the AMHC are aggregated over all S and G-Experts and include the experts' self weights. Reference should be made to Section 2 and Appendix C of Vol. 1 for a discussion on these estimators. Briefly, in our elicitation process we asked each S-expert to indicate which set of zones he considered his "best estimate" in the sense that it represented the most likely of all of his choices and similarly for the best estimate values for all of the seismicity parameters for each zone. We also asked each G-Expert to indicate which ground motion model represented his best estimate model. Then, as

indicated in Vol. I, the set of best estimate zones and seismicity parameters are used with each of the best estimate ground motion models to generate 55 BEHCs' (i.e. 11 S-Experts X 5 G-Experts). These 55 curves are then aggregated using both the S and G-Experts' self weights. The AMHC curves are generated in the usual manner using all 2750 simulations of the Monte Carlo analysis.

Figures 2.SN.2 give the BEHC for each S-Expert aggregated over the five G-Experts. Whenever individual S-Experts' hazard curves are plotted they are denoted by the plot key given in Table 2.0. Figures 2.SN.2 give a measure of the range of difference of opinion between the S-Experts.

Figures 2.SN.3 give the median and a measure of the overall uncertainty in the form of the 15th, 50th and 85th constant percentile hazard curves (CPHCs) based on all 2750 simulations.

Figures 2.SN.4 give the contribution to the BEHC (aggregated over all S and G-Experts) for earthquakes in four magnitude ranges:

<u>Curve Number</u> <u>Magnitude Range</u>	
1	$3.75 \leq m_b \leq 5$
2	$5 \leq m_b \leq 5.75$
3	$5.75 \leq m_b \leq 6.5$
4	$6.5 \leq m_b$

The curves of Figures 2.SN.4 are useful to indicate the relative contribution of smaller, moderate and large earthquakes to the seismic hazard and how much higher the estimated seismic hazard would be if the contribution of smaller earthquakes in the range 3.75 to 5.0 were included.

Figures 2.SN.5 give the best estimate uniform hazard spectra (BEUHS) for return periods of 500,1000,2000,5000, and 10,000 years, aggregated over all S and G-Experts.

Figures 2.SN.6 give the 1000 year return period BEUHS for each of the S-Experts, aggregated over the G-Experts. The S-Experts' BEUHS are plotted using the symbols in Table 2.0. These plots give a good measure of the significance of the differences in opinion between the S-Experts.

Figures 2.SN.7,8,9 give the 15th, 50th and 85th constant percentile uniform hazard spectra (CPUHS) aggregated over all S and G-Experts for return periods of 500,1000 and 10,000 years. The spread between the 15th and 85th CPUHS gives a good measure of the overall uncertainty in the estimate of the seismic hazard at the site.

Figures 2.SN.10 give the 50th CPUHS for return periods of 500,1000,2000, 5000 and 10,000 years, aggregated over all S and G-Experts.

Discussion will only be given if some factors of interest are noted. In Section 3 comparisons between the sites and general observations are made.

It should be noted that all the sites considered in this report (Batch 1 sites) listed in Table 1.1 were assumed to be in the Region 1 (see Fig. 2.3 of Vol 1). This region of North America, which we refer to as Region 1 in some of the figures in this report, and which roughly encompassed the North East of the United States and parts of Eastern Canada, is _not_ related to the Region 1 as it is defined by the NRC, for regulation purposes. The Hope Creek, Limerick, Peach Bottom, Salem, Susquehanna and Three Mile Island sites are very near the border and could be placed in other regions. It only applies for Region 1 as G-Expert 2 specified different BE GM models for Region 1 than in Regions 2-4. The significance of this assumption is discussed in Section 3.

TABLE 2.0

PLOT SYMBOL KEY USED FOR INDIVIDUAL S-EXPERTS ON FIGS. 2.SN.2 and 2.SN.6

Expert No.	Plot Symbol
1	1
2	2
3	3
4	4
5	5
6	6
7	7
10	A
11	B
12	C
13	D

2.1 FITZPATRICK

Fitzpatrick is a rock site and is represented by the symbol "1" in Fig. 1.1. Table 2.1.1 and Figs. 2.1.1 to Figs. 2.1.10 give the basic results for the Fitzpatrick site. The large spread between the BEHC and AMHC in Fig. 2.1.1 indicates that there is considerable diversity of opinion between the experts with some relatively high and low outliers. The AMHC is close to the 85th percentile CPHC.

The fact that the 85th percentile CPHC is so much higher than the 50th CPHC suggests that the distribution of hazard curves is bimodal. This is discussed later in Section 3.0 and is in part explained by the presence of a high ground motion model, as discussed in detail in Section 3.2 of Volume VI.

Figure 2.1.4 indicates that most of the hazard is coming from earthquakes in the $5.0 \leq m_b \leq 6.5$ range, with the range $5.75 \leq m_b \leq 6.5$ being the most significant. Figure 2.1.4 also suggests that the hazard curve would only change for PGA values less than 0.15g if earthquakes in the range $3.75 \leq m_b \leq 5$ were to be included.

Examination of Table 2.1.1 reveals several interesting results about which zones are contributing most to the hazard the Fitzpatrick site. In particular for S-Experts 1,2,4,5,11,12 and 13 somewhat distant zones are significant. There are several reasons for this. First, it should be recalled that the percent contributed listed in Table 2.1.1 is relative to the BE map. Thus, for example, if we examine the map for S-Expert 2 in Appendix B, we see that zones 33 and 26 are very near the site, yet they are not listed in Table 2.1.1. This is because zones 26 and 33 for S-Expert 2 only have a probability of existence of only 0.4 and thus are not part of the BE map (see Table B2.1 in Appendix B). It is likely that zone 33 contributes significantly to the uncertainty at the Fitzpatrick site. Secondly, the BEHC are aggregated over the G-Experts (Per S-Expert) arithmetically so that a high outlier tends to dominate the results. It is shown in Section 3.2 of Volume VI how G-Expert 5's ground motion model leads to BEHCs that are high outliers (relative to the other G-Experts' BEHCs per S-Expert). G-Expert 5's BEHCs are high in the sense indicated for several reasons:

(1) Most importantly, as discussed in Vol. 1 Section 3.5, for the same distance and magnitude, the Model G16-A3 (G-Expert 5's choice for BE GM model) is higher by a factor of 2 relative to the other BE GM models for rock sites. A factor of 2 in PGA results approximately a factor of 8-10 in probability of exceedance. (See also Volume VI Section 3.2.)

(2) It can be seen from Fig. 3.4 in Vol. 1 that G-Expert 5's BE PGA (GM model G16-A3) has significantly lower attenuation than the other models particularly at the larger magnitudes. This coupled with the

site correction factor for rock increases the contribution from distant zones which have larger earthquakes. For example, a simple calculation would show that earthquakes of $m_b = 6.0$ have the same PGA at 100 km as $m_b = 5$ earthquakes at 20 km.

(3) G-expert 5 set the random uncertainty (standard deviation on the natural log of the PGA) to 0.7 compared to the range of values (0.35 - 0.55) selected by the other G-Experts. Relative to results obtained with a value of 0.55, this larger uncertainty (0.7) leads to an increase in the G-Expert 5's BEHC by about a factor of 2 in probability of exceedance at lower (0.2g) g-values to over a factor of 3 at high g-values (0.9g).

In summary we typically expect at rock sites that BEHC for G-Expert 5 for any S-Expert will be about a factor of 10-20 higher in probability of exceedance relative to the other BE GM models (factors (1) and (3) noted above) as illustrated in Fig. 2.1.11. In Fig. 2.1.11 we plot the BEHCs per G-Expert for S-Expert 3's seismicity input. At the Fitzpatrick site we have for some S-Experts the case where the zone that contains the site has a low upper magnitude cutoff and a nearby zone which has a significantly larger upper magnitude cutoff. In this case the low attenuation of model G16-A3 becomes very important and G-Expert 5's BEHC becomes significantly larger (relative to the typical case illustrated in Fig. 2.1.11) than the other BEHCs per G-Expert for a given S-Expert's input. Such a case is shown in Fig. 2.1.12 where the BEHC per G-Expert based on S-Expert 1's input is plotted. It can be seen from Table 2.1.1 and the map for S-Expert 1 in Appendix B zone 20 which contribute most to the hazard is some distance from the site. The upper magnitude cutoff in zone 15 which contains the site is only 5.8 whereas it is 6.7 for zone 20. In addition, based on the parameters given in Appendix B for S-Expert 1, it is easy to show that the rate of earthquakes per unit area in the magnitude 5-6 range is over 100 times higher in zone 20 than in zone 15, thus the BEHC for G-Expert 5 is much higher than the other BEHC. It should be noted that other GM models will result in hazard curves where the more distant zones are important. In particular the models SE-1A and COMB-1a are two such models. Additional discussion is given in Section 3.

TABLE 2.1.1

MOST IMPORTANT ZONES PER S-EXPERT
FOR FITZPATRICK

SITE SOIL CATEGORY ROCK

S-XPT NUM.	MOST ZONE		ZONES CONTRIBUTING MOST SIGNIFICANTLY TO THE PGA BEHC AND % OF CONTRIBUTION AT LOW PGA(0.125G)				AT HIGH PGA(0.60G)			
1	ZONE 15	ZONE ID:	ZONE 20	ZONE 21	ZONE 19	ZONE 4	ZONE 20	ZONE 19	ZONE 21	ZONE 15
		% CONT.:	71.	11.	9.	4.	85.	9.	4.	2.
2	COMP. ZO	ZONE ID:	ZONE 32	ZONE 31	COMP. ZON	ZONE 28	ZONE 32	COMP. ZON	ZONE 31	ZONE 28
		% CONT.:	66.	17.	14.	3.	74.	23.	3.	0.
3	ZONE 11	ZONE ID:	ZONE 11	ZONE 2	ZONE 3	ZONE 4	ZONE 11	ZONE 2	ZONE 3	COMP. ZON
		% CONT.:	63.	27.	5.	2.	83.	16.	0.	0.
4	ZONE 13	ZONE ID:	ZONE 16	ZONE 19	ZONE 15	ZONE 18	ZONE 16	ZONE 13	ZONE 19	ZONE 15
		% CONT.:	78.	7.	4.	4.	98.	1.	0.	0.
5	COMP. ZO	ZONE ID:	ZONE 6	ZONE 4	ZONE 5	ZONE 3	ZONE 6	ZONE 5	ZONE 4	ZONE 3
		% CONT.:	96.	2.	1.	1.	95.	2.	2.	1.
6	COMP. ZO	ZONE ID:	ZONE 7	ZONE 3	ZONE 8	ZONE 6	ZONE 7	COMP. ZON	ZONE 3	ZONE 8
		% CONT.:	75.	13.	5.	3.	95.	2.	2.	1.
7	ZONE 41	ZONE ID:	ZONE 17	ZONE 18	ZONE 41	ZONE 26	ZONE 17	ZONE 41	ZONE 18	ZONE 26
		% CONT.:	56.	24.	11.	5.	70.	19.	10.	1.
10	ZONE 19	ZONE ID:	ZONE 19 =	ZONE 6	ZONE 7	ZONE 9	ZONE 19 =	ZONE 6	ZONE 7	ZONE 9
		% CONT.:	32.	30.	20.	8.	67.	26.	5.	2.
11	CZ = ZON	ZONE ID:	ZONE 3	ZONE 4	CZ = ZONE	ZONE 2	ZONE 4	CZ = ZONE	ZONE 3	ZONE 9
		% CONT.:	43.	19.	18.	8.	47.	27.	23.	2.
12	ZONE 4 =	ZONE ID:	ZONE 31	ZONE 30	ZONE 34	ZONE 31A	ZONE 31	ZONE 30	ZONE 34	ZONE 31A
		% CONT.:	72.	15.	9.	2.	76.	19.	6.	0.
13	CZ 15	ZONE ID:	ZONE 11	CZ 15	ZONE 12	ZONE 10	CZ 15	ZONE 11	ZONE 12	ZONE 7
		% CONT.:	41.	34.	22.	3.	83.	12.	5.	0.

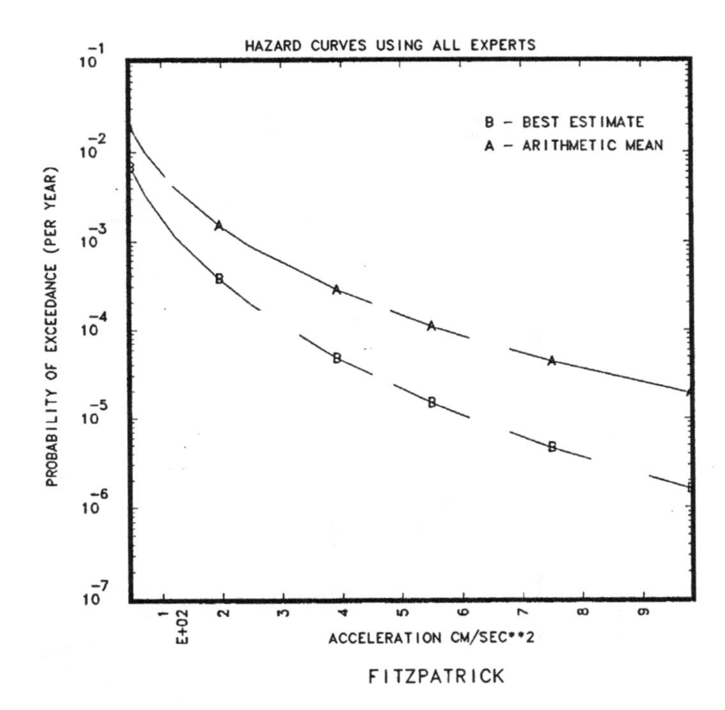

Figure 2.1.1 Comparison of the BEHC and AMHC aggregated over all S and G-Experts for the Fitzpatrick site.

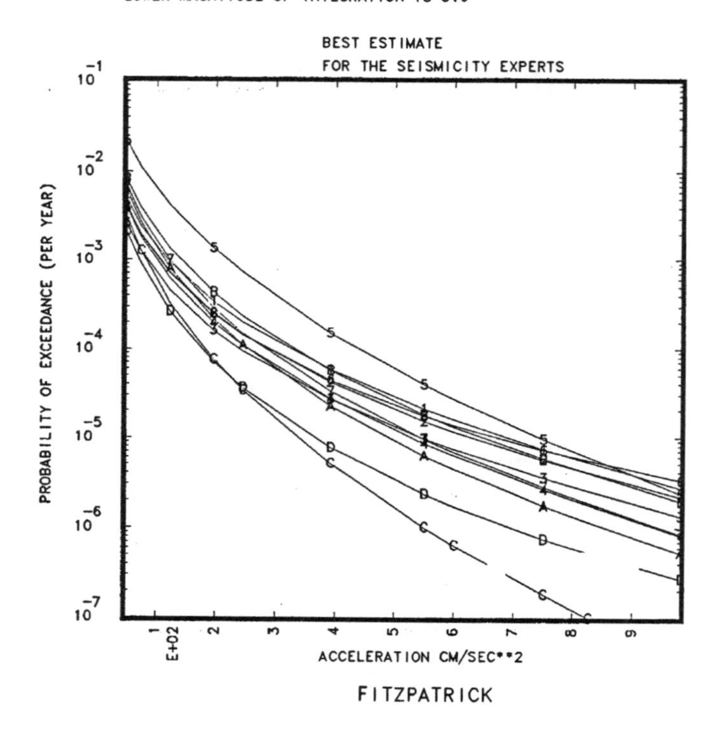

Figure 2.1.2 BEHCs per S-Expert combined over all G-Experts for the Fitzpatrick site. Plot symbols given in Table 2.0.

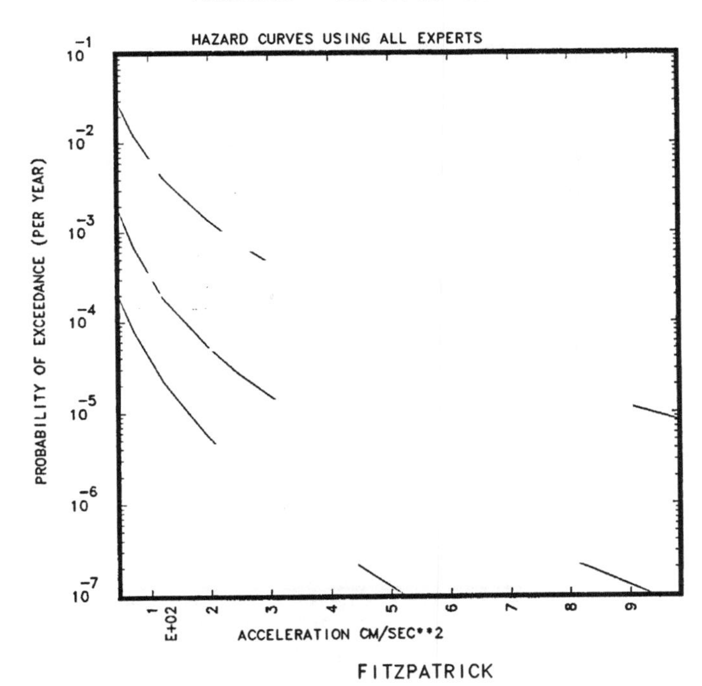

Figure 2.1.3 CPHCs for the 15th, 50th and 85th percentiles based on all S and G-Experts' input for the Fitzpatrick site.

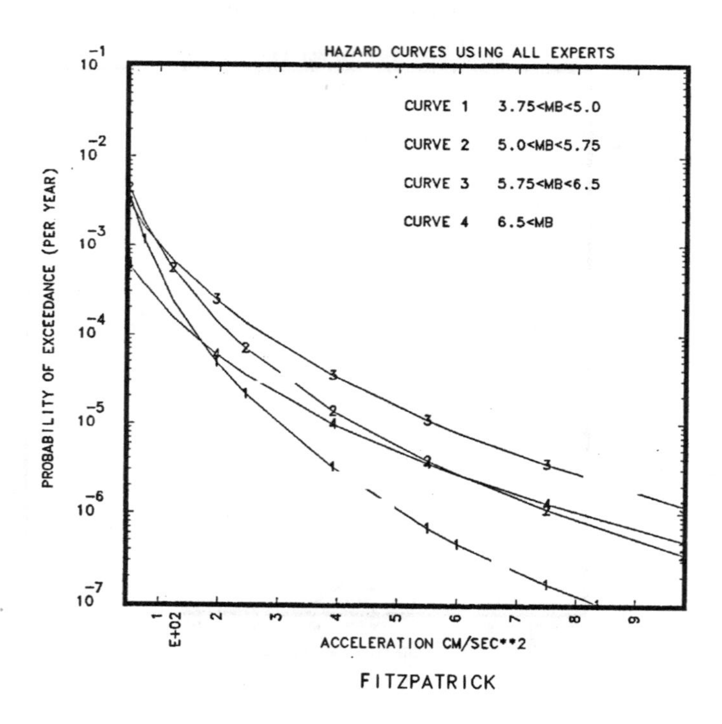

Figure 2.1.4 BEHCs which include only the contribution to the PGA hazard from earthquakes within the indicated magnitude range for the Fitzpatrick site.

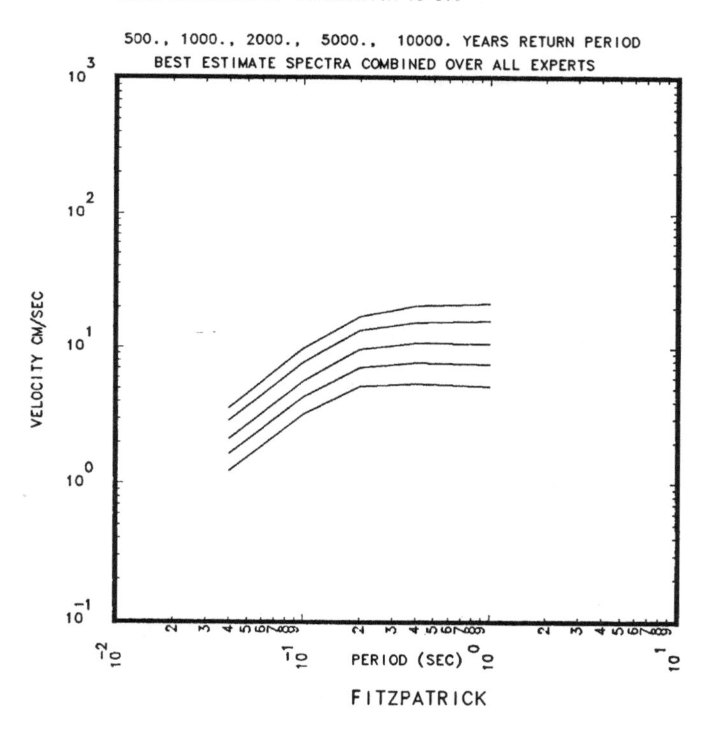

Figure 2.1.5 BEUHS for return periods of 500, 1000, 2000, 5000 and 10000 years aggregated over all S and G-Experts for the Fitzpatrick site.

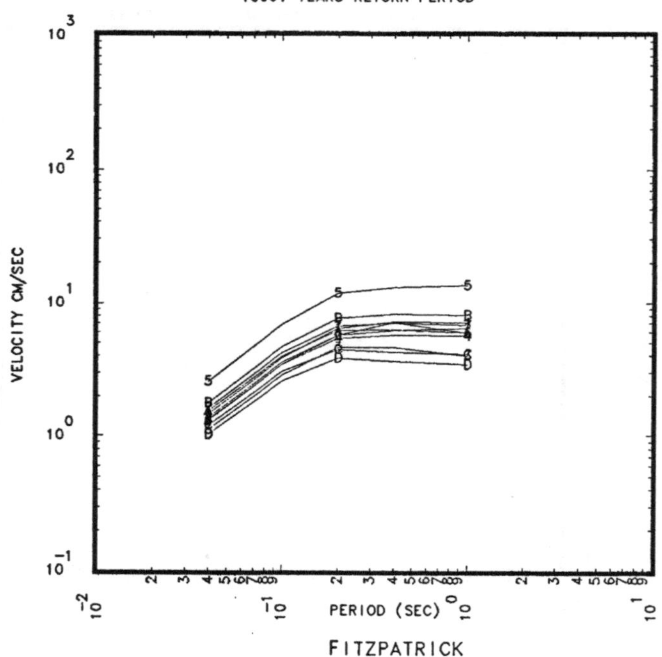

Figure 2.1.6 The 1000 year return period BEUHS per S-Expert aggregated over all G-Experts for the Fitzpatrick site. Plot symbols are given in Table 2.0.

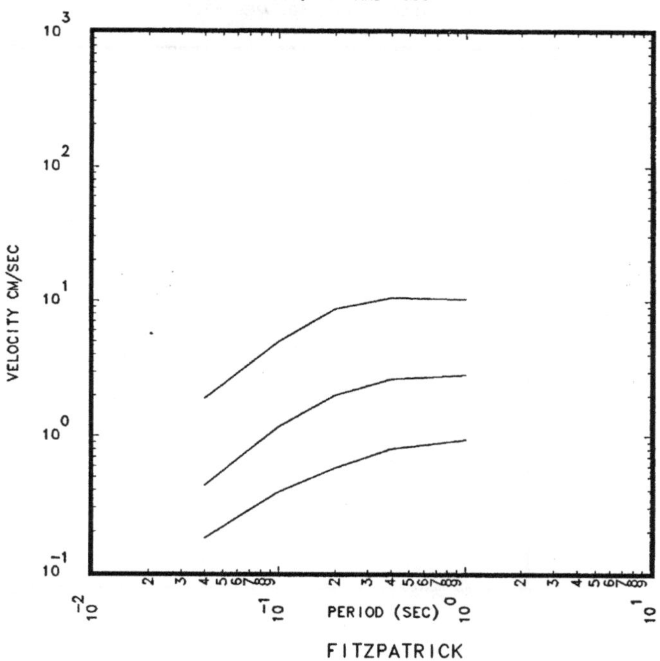

Figure 2.1.7 500 year return period CPUHS for the 15th, 50th and 85th percentiles aggregated over all S and G-Experts for the Fitzpatrick site.

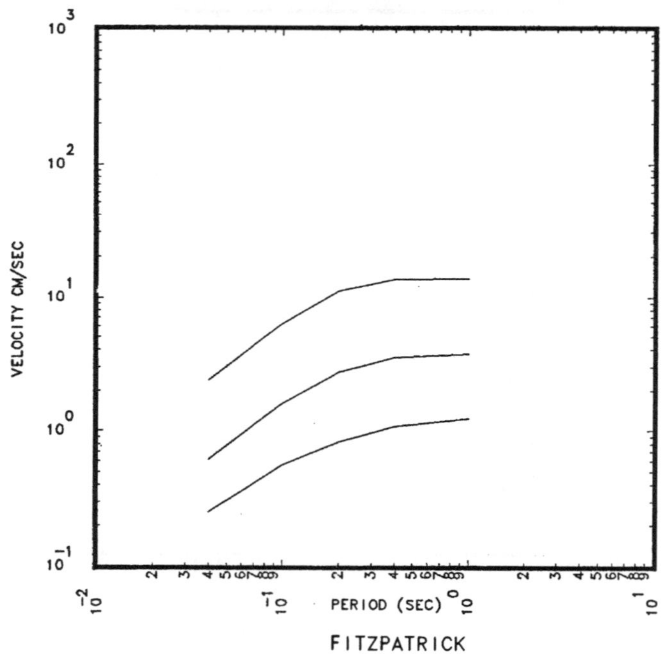

Figure 2.1.8 1000 year return period CPUHS for the 15th, 50th and 85th percentile aggregated over all S and G-Experts for the Fitzpatrick site.

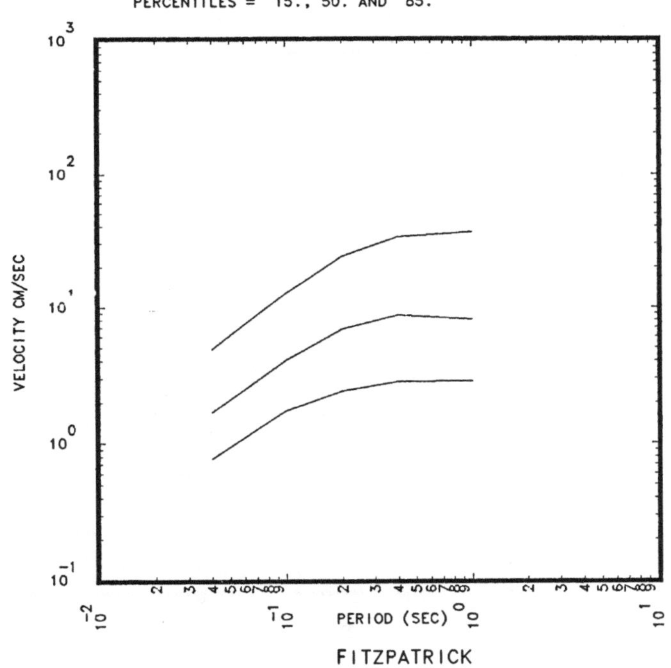

Figure 2.1.9 10000 year return period CPUHS for the 15th, 50th and 85th percentiles aggregated over all S and G-Experts for the Fitzpatrick site.

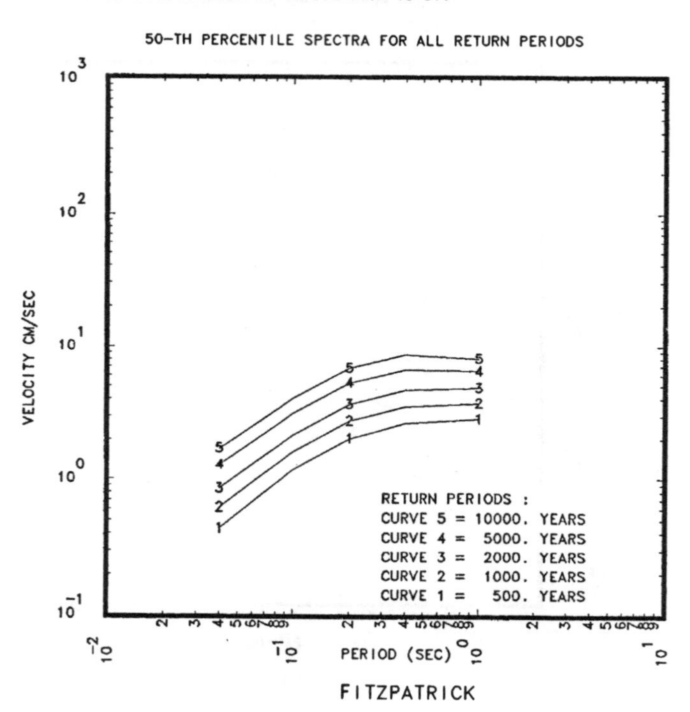

Figure 2.1.10 Comparison of the 50th percentile CPUHS for return periods of 500, 1000, 2000, 5000 and 10000 years for the Fitzpatrick site.

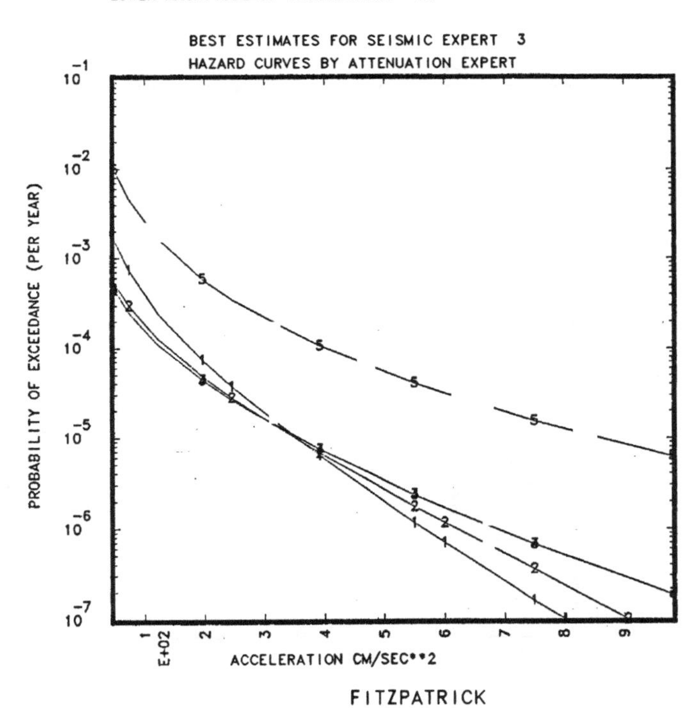

Figure 2.1.11. Comparison of the BEHCs for PGA per G-Expert for S-Expert 3's input for the Fitzpatrick site. The spread between G-Expert 5's BEHC and the BEHCs for the other G-Experts' BEHCs is typical for rock sites located in region 1.

EUS SEISMIC HAZARD CHARACTERIZATION, SEPT. 1987
LOWER MAGNITUDE OF INTEGRATION = 5.

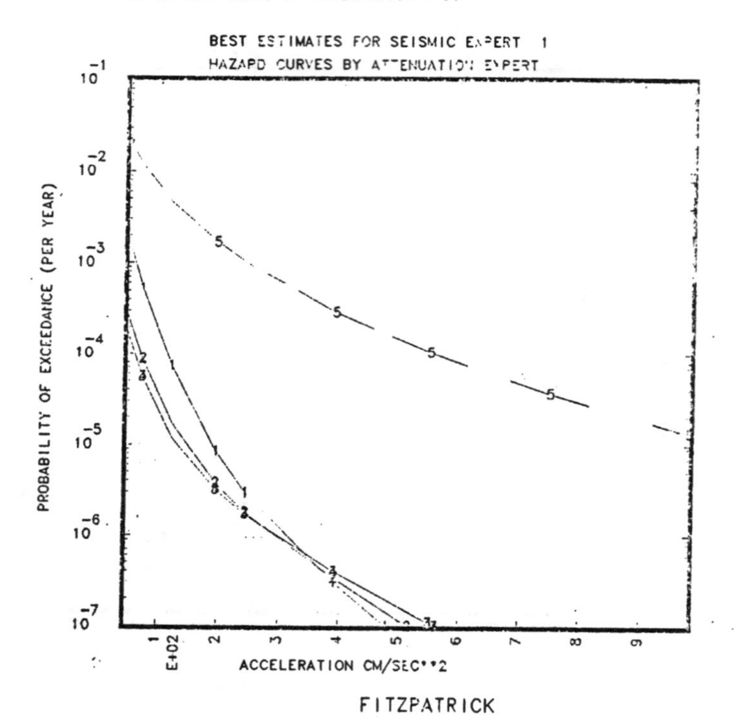

Figure 2.1.12. Comparison of the BEHCs for PGA per G-Expert for S-Expert 1's input for the Fitzpatrick site. The spread between G-Expert 5's BEHC and the other G-Experts' BEHCs is only typical for S-Experts 1, 2, 4, 5, 11, 12, and 13 for the Fitzpatrick, Ginna, and Nine Mile Point sites.

2.2 GINNA

Ginna is a rock site and is represented by symbol "2"in Fig. 1.1. Table 2.2.1 and Figs. 2.2.1 to 2.2.10 give the basic results for the Ginna site. The large spread between the BEHC and AMHC indicates a large diversity of opinion among both the S and G-Experts. The AMHC generally lies above the 85th percentile CPHC indicating some high outliers.

The BEHC is also very high, in part, because, as can be seen from Fig. 2.2.2, S-Expert 5's BEHC is much higher than the other S-Experts. This occurs because only S-Expert 5 had the Ginna site in a relatively small zone with relatively high seismicity. Also, zone 6 is an extremely important zone for Expert 5. It is in fact the contribution from zone 6 which dominates the BEHC. However the probability of existence of zone 6 is only 0.5. Thus it only occurs in one half of the maps. For this case, the contributions given in Table 2.2.1 for S-Expert 5 are a bit misleading as this zone's (zone 6) probability of existence is not factored in. Zone 26 of S-Expert 2 had a zone similar to S-Expert 5's zone 5, however the probability of existence of zone 26 is only 0.4. Thus as indicated in Section 2.0, S Expert 2's zone 26 is not included in the tabulation given in Table 2.2.1 because only the best estimate zones are included in Table 2.2.1. For such cases as this the BEHC and listing of zonal dominance tend to give a somewhat distorted picture of the seismic hazard.

Figure 2.2.4 indicates that most of the hazard is coming from earthquakes in the $5.0 \leq m_b \leq 6.5$ range with the range $5.75 \leq m_b \leq 6.5$ being most important. It is also seen from Fig. 2.2.4 that inclusion of earthquakes in the range $3.75 \leq m_b \leq 5.0$ would only have much impact on the BEHC for PGA values less than 0.15g. The conclusions drawn here are based on BEHC, thus do not account for all the zones and ground motion models and do not account for the modeling uncertainty. However, in tests performed all along this study, we verified that those conclusions were correct. This is due mostly to the fact that the BE case strongly dominates the results.

The discussion given in Section 2.1 relative to the dominance of G-Expert 5's GM model for the BEHC and AMHC also applies.

TABLE 2.2.1

MOST IMPORTANT ZONES PER S-EXPERT
FOR GINNA

SITE SOIL CATEGORY ROCK

S-XPT NUM.	HOST ZONE		ZONES CONTRIBUTING MOST SIGNIFICANTLY TO THE BA BEHC AND % OF CONTRIBUTION AT LOW PGA(0.125G)				AT HIGH PGA(0.60G)			
1	ZONE 15	ZONE ID:	ZONE 20	ZONE 19	ZONE 21	ZONE 4	ZONE 19	ZONE 20	ZONE 21	ZONE 15
		% CONT.:	51.	25.	11.	8.	56.	37.	3.	3.
2	[illegible]. ZO	ZONE ID:	ZONE 32	COMP. ZON	ZONE 31	ZONE 28	ZONE 32	COMP. ZON	ZONE 31	ZONE 28
		% CONT.:	62.	19.	12.	5.	61.	38.	1.	0.
3	ZONE 11	ZONE ID:	ZONE 11	ZONE 2	ZONE 3	ZONE 5	ZONE 1	ZONE 2	ZONE 3	COMP. ZON
		% CONT.:	78.	15.	4.	2.	[illegible].	3.	0.	0.
4	ZONE 13	ZONE ID:	ZONE 16	ZONE 19	ZONE 14	ZONE 12	ZONE 16	ZONE 13	ZONE 14	ZONE 19
		% CONT.:	75.	8.	6.	3.	96.	3.	0.	0.
5	ZONE 5	ZONE ID:	ZONE 6	ZONE 5	ZONE 4	ZONE 3	ZONE 6	ZONE 5	ZONE 3	ZONE 4
		% CONT.:	97.	3.	0.	0.	95.	4.	0.	0.
6	ZONE 8	ZONE ID:	ZONE 7	ZONE 8	ZONE 3	ZONE 6	ZONE 8	ZONE 7	COMP. ZON	ZONE 3
		% CONT.:	46.	36.	10.	3.	57.	37.	4.	1.
7	ZONE 41	ZONE ID:	ZONE 17	ZONE 41	ZONE 12	ZONE 18	ZONE 41	ZONE 7	ZONE 12	ZONE 18
		% CONT.:	43.	16.	16.	15.	46.	34.	14.	4.
10	ZONE 19	ZONE ID:	ZONE 19 =	ZONE 9	ZONE 7	ZONE 6	ZONE 19 =	ZONE 9	ZONE 6	ZONE 7
		% CONT.:	34.	34.	13.	11.	53.	43.	2.	1.
11	CZ = ZON	ZONE ID:	ZONE 3	ZONE 9	CZ = ZONE	ZONE 4	ZONE 9	CZ = ZONE	ZONE 4	ZONE 3
		% CONT.:	33.	24.	18.	13.	36.	30.	20.	11.
12	ZONE 4 =	ZONE ID:	ZONE 31	ZONE 34	ZONE 30	ZONE 31A	ZONE 31	ZONE 34	ZONE 30	ZONE 31A
		% CONT.:	79.	10.	5.	4.	91.	9.	0.	0.
13	CZ 15	ZONE ID:	CZ 15	ZONE 11	ZONE 12	ZONE 10	CZ 15	ZONE 12	ZONE [illegible]	ZONE 7
		% CONT.:	48.	29.	20.	2.	95.	2.	2.	0.

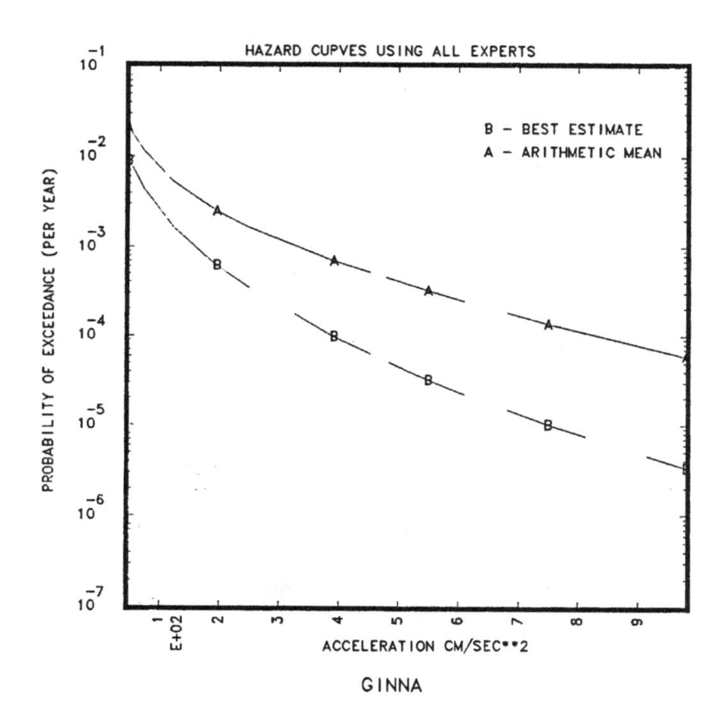

Figure 2.2.1 Comparison of the BEHC and AMHC aggregated over all S and G-Experts for the Ginna site.

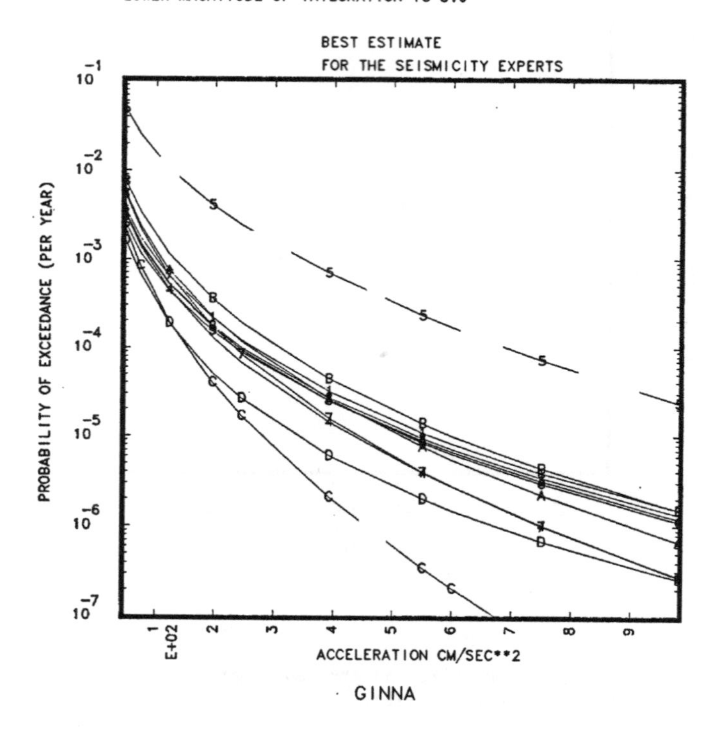

Figure 2.2.2 BEHCs per S-Expert combined over all G-Experts for the Ginna site. Plot symbols given in Table 2.0.

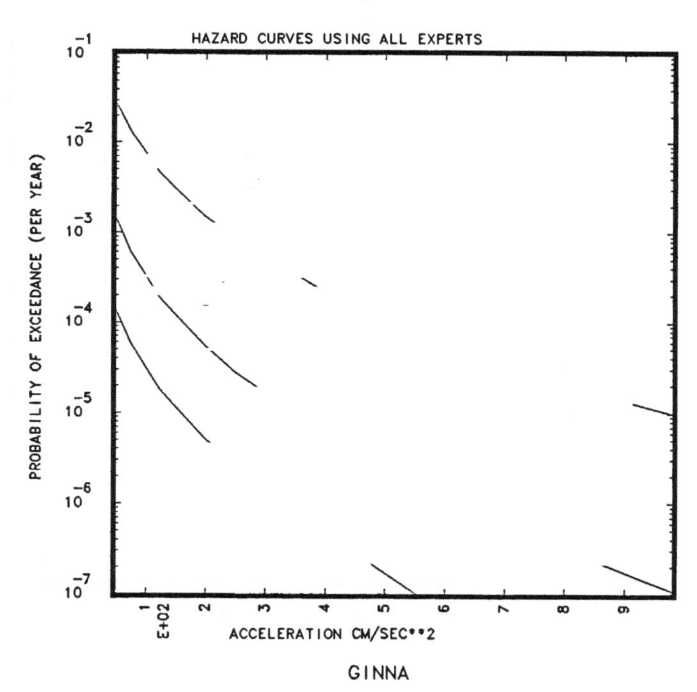

Figure 2.2.3 CPHCs for the 15th, 50th and 85th percentiles based on all S and G-Experts' input for the Ginna site.

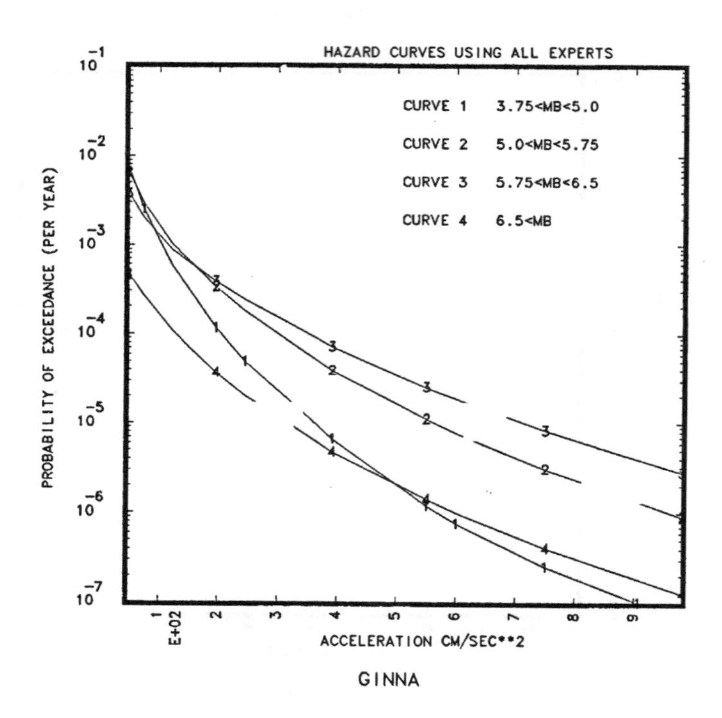

Figure 2.2.4 BEHCs which include only the contribution to the PGA hazard from earthquakes within the indicated magnitude range for the Ginna site.

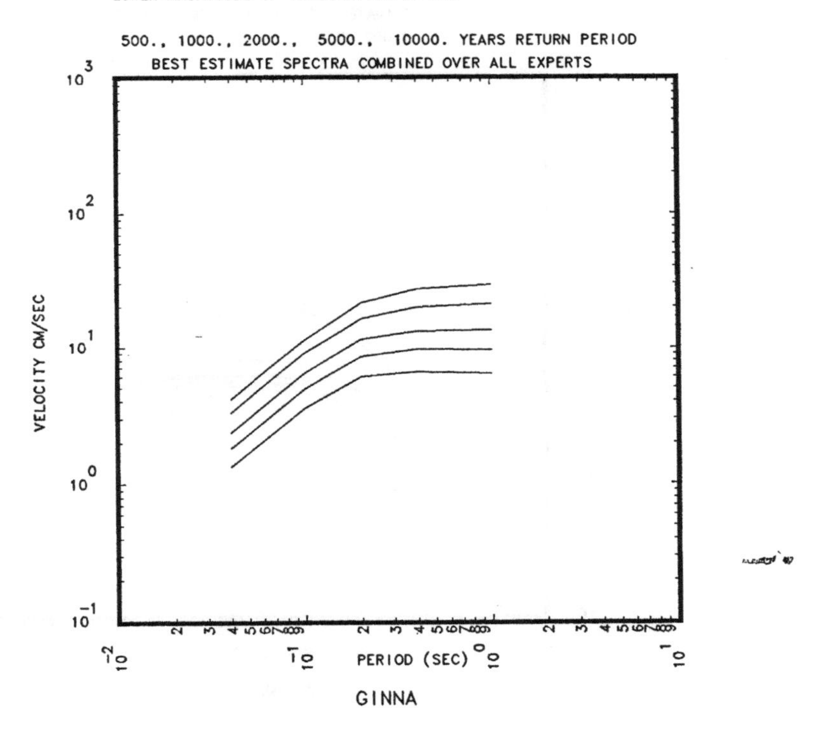

Figure 2.2.5 BEUHS for return periods of 500, 1000, 2000, 5000 and 10000 years aggregated over all S and G-Experts for the Ginna site.

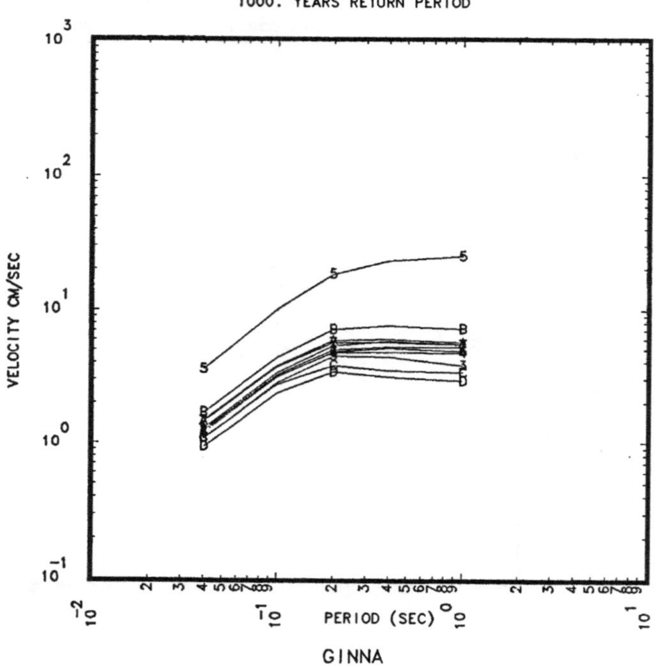

Figure 2.2.6 The 1000 year return period BEUHS per S-Expert aggregated over all G-Experts for the Ginna site. Plot symbols are given in Table 2.0.

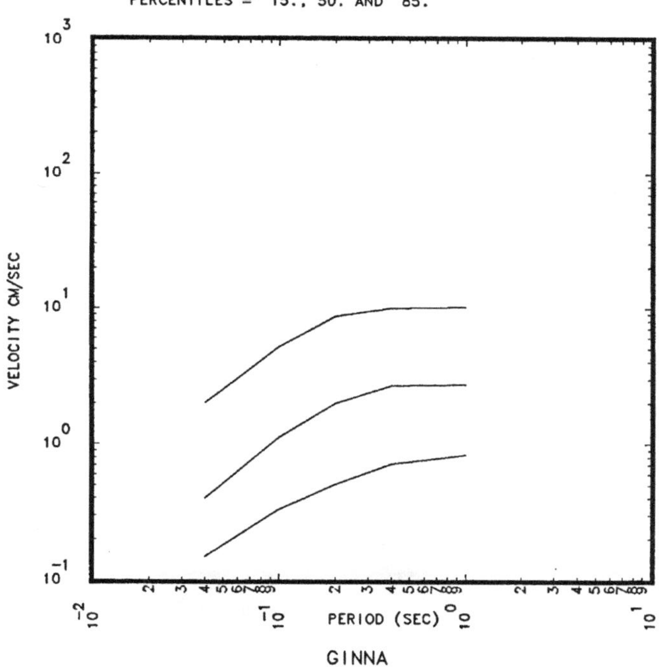

Figure 2.2.7 500 year return period CPUHS for the 15th, 50th and 85th percentiles aggregated over all S and G-Experts for the Ginna site.

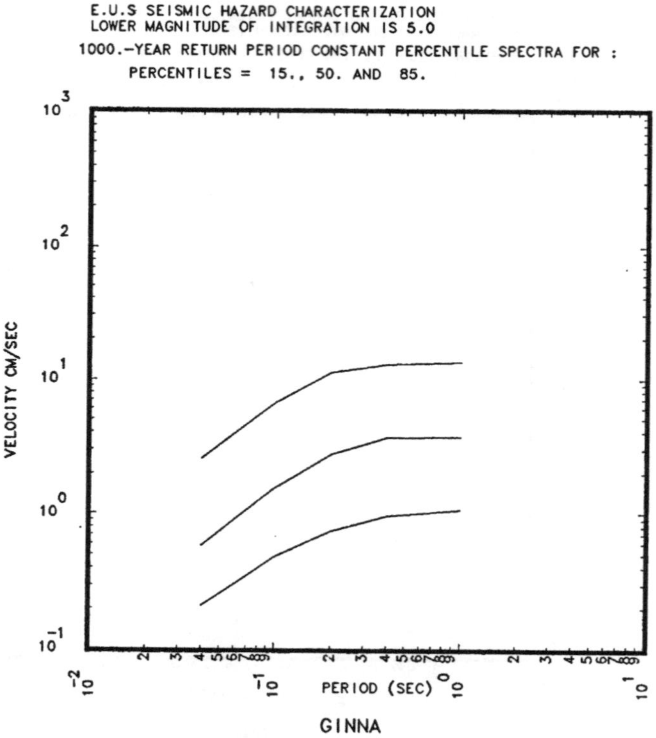

E.U.S SEISMIC HAZARD CHARACTERIZATION
LOWER MAGNITUDE OF INTEGRATION IS 5.0
1000.-YEAR RETURN PERIOD CONSTANT PERCENTILE SPECTRA FOR :
PERCENTILES = 15., 50. AND 85.

VELOCITY CM/SEC

PERIOD (SEC)

GINNA

Figure 2.2.8 1000 year return period CPUHS for the 15th, 50th and 85th percentile aggregated over all S and G-Experts for the Ginna site.

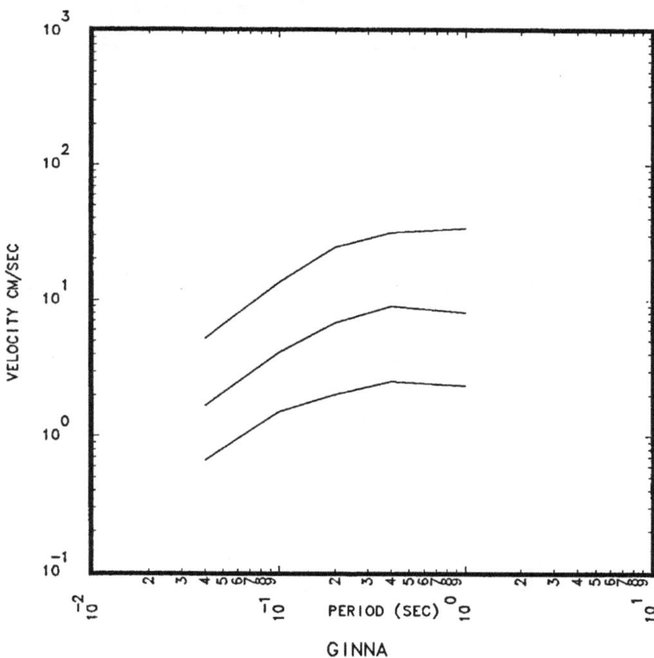

Figure 2.2.9 10000 year return period CPUHS for the 15th, 50th and 85th percentiles aggregated over all S and G-Experts for the Ginna site.

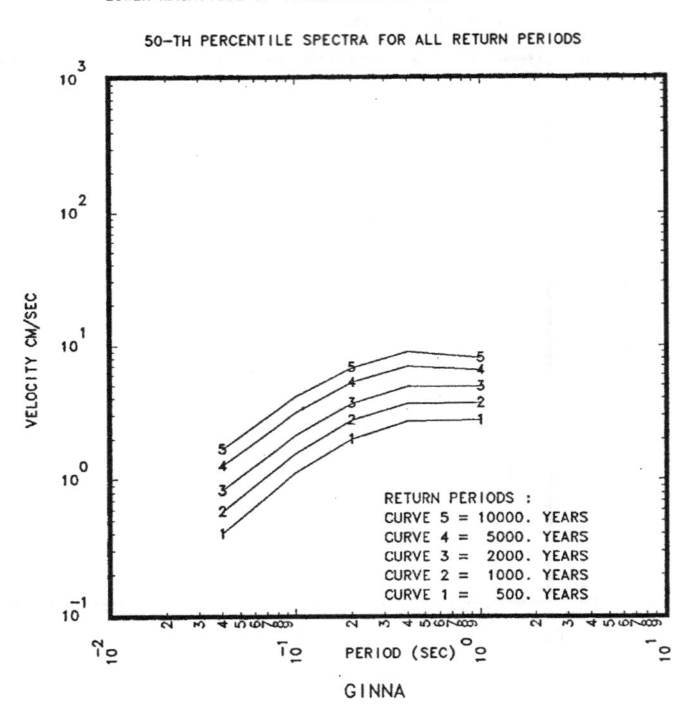

Figure 2.2.10 Comparison of the 50th percentile CPUHS for return periods of 500, 1000, 2000, 5000 and 10000 years for the Ginna site.

2.3 HADDAM NECK

Haddam Neck is a rock site and is represented by the symbol "3" in Fig. 1.1. Table 2.3.1 and Figs. 2.3.1 to 2.3.10 give the basic results for the Haddam Neck site. The AMHC is close to the 85th percentile CPHC.

We see from Table 2.3.1 that for most S-Experts the zone which contributes most to the BEHC is the zone that contains the site. Usually this means that the relative spread between BEHCs per G-Expert for any given S-Expert is limited because the spreading effect due to G-Expert 5 does not show up in local seismicity. Thus there is much less spread between the BEHC and AMHC as compared to Fitzpatrick and Ginna for which large and distant earthquakes contribute significantly to the overall hazard, and thus the spreading effect of G-Expert 5 becomes important. In addition we would expect that the contribution of small earthquakes to the BEHC would be more significant than when the contribution of some distant zone was more significant than the local zone containing the site.

We see from Fig. 2.3.4 that this is the case for the Haddam Neck site because if earthquakes in the 3.75 to 5.0 range were included, the BEHC would be about a factor of 2 higher at 0.05g, a factor of 1.3 higher at about 0.2g and decreasing to nearly a factor of 1.0 at about 0.6g.

TABLE 2.3.1

MOST IMPORTANT ZONES PER S-EXPERT
FOR HADDAM NECK

SITE SOIL CATEGORY ROCK

S-XPT NUM.	HOST ZONE		ZONES CONTRIBUTING MOST SIGNIFICANTLY TO THE PGA BEHC AND % OF CONTRIBUTION AT LOW PGA(0.125G)				AT HIGH PGA(0.60G)			
1	ZONE 22	ZONE ID:	ZONE 22	ZONE 20	ZONE 21	ZONE 4	ZONE 22	ZONE 20	ZONE 1	ZONE 21
		% CONT.:	56.	24.	12.	5.	88.	6.	3.	2.
2	ZONE 31	ZONE ID:	ZONE 31	ZONE 32	ZONE 28	COMP. ZON	ZONE 31	ZONE 32	COMP. ZON	ZONE 28
		% CON T:	68.	23.	7.	2.	87.	12.	1.	0.
3	ZONE 4	ZONE ID:	ZONE 4	ZONE 2	ZONE 3	ZONE 5	ZONE 4	ZONE 2	ZONE 5	ZONE 3
		% CON T:	87.	4.	4.	4.	9.	0.	0.	0.
4	ZONE 23	ZONE ID:	ZONE 23	ZONE 18	ZONE 16	ZONE 19	ZONE 23	ZONE 18	ZONE 16	ZONE 11
		% CON T:	31.	27.	16.	8.	75.	12.	6.	4.
5	ZONE 1	ZONE ID:	ZONE 1	ZONE 6	ZONE 3	ZONE 8	ZONE 1	ZONE 3	ZONE 8	ZONE 4
		% CONT.:	78.	14.	5.	2.	99.	1.	0.	0.
6	ZONE 6	ZONE ID:	ZONE 6	ZONE 7	ZONE 3	ZONE 5	ZONE 6	ZONE 7	ZONE 5	ZONE 3
		% CON T.:	73.	10.	9.	8.	93.	3.	3.	1.
7	ZONE 15	ZONE ID:	ZONE 19	ZONE 17	ZONE 15	ZONE 14	ZONE 15	ZONE 19	ZONE 24	ZONE 17
		% CONT.:	28.	14.	14.	9.	38.	27.	13.	6.
10	ZONE 4A	ZONE ID:	ZONE 2	ZONE 4A	ZONE 22	ZONE 23	ZONE 4A	ZONE 2	ZONE 22	ZONE 19
		% CON T:	23.	22.	11.	10.	69.	19.	6.	2.
11	CZ = ZON	ZONE ID:	ZONE 3	ZONE 5	ZONE 1	CZ = ZONE	ZONE 5	CZ = ZONE	ZONE 1	ZONE 3
		% CON T:	33.	27.	19.	11.	48.	19.	17.	11.
12	ZONE 32	ZONE ID:	ZONE 32	ZONE 31	ZONE 34	ZONE 33	ZONE 32	ZONE 34	ZONE 31	ZONE 33
		% CON T.:	72.	11.	9.	5.	99.	1.	0.	0.
13	ZONE 10	ZONE ID:	ZONE 10	ZONE 12	CZ 15	ZONE 11	ZONE 10	CZ 15	ZONE 12	ZONE 11
		% CON T:	88.	7.	2.	2.	9.	0.	0.	0.

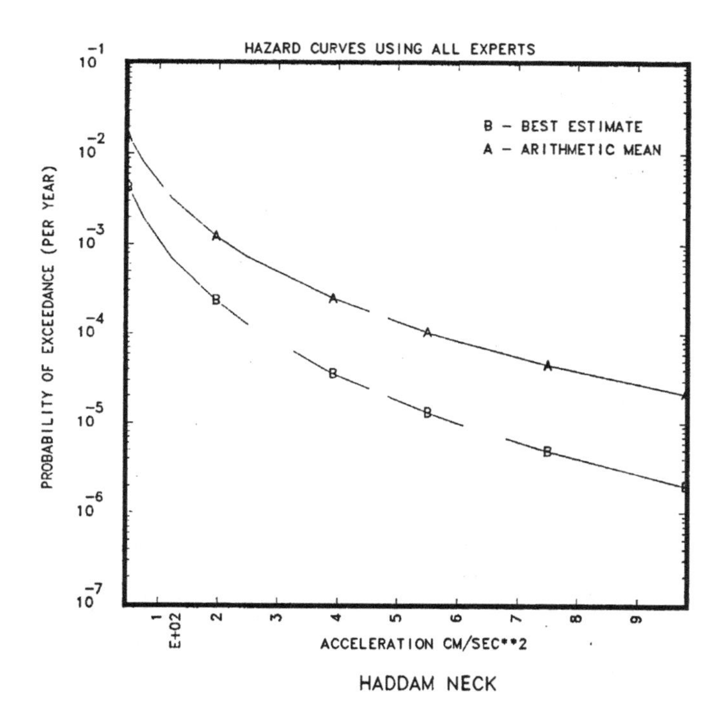

Figure 2.3.1 Comparison of the BEHC and AMHC aggregated over all S and G-Experts for the Haddam Neck site.

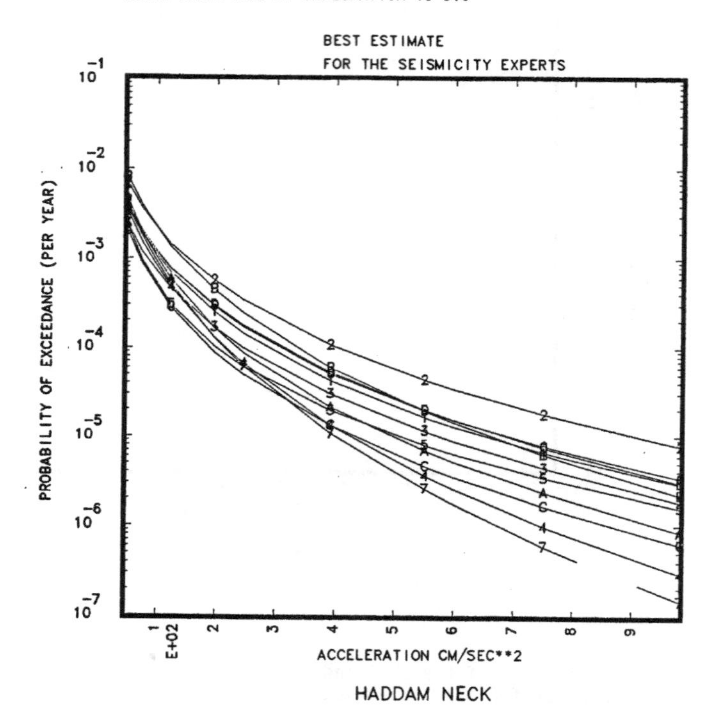

Figure 2.3.2 BEHCs per S-Expert combined over all G-Experts for the Haddam Neck site. Plot symbols given in Table 2.0.

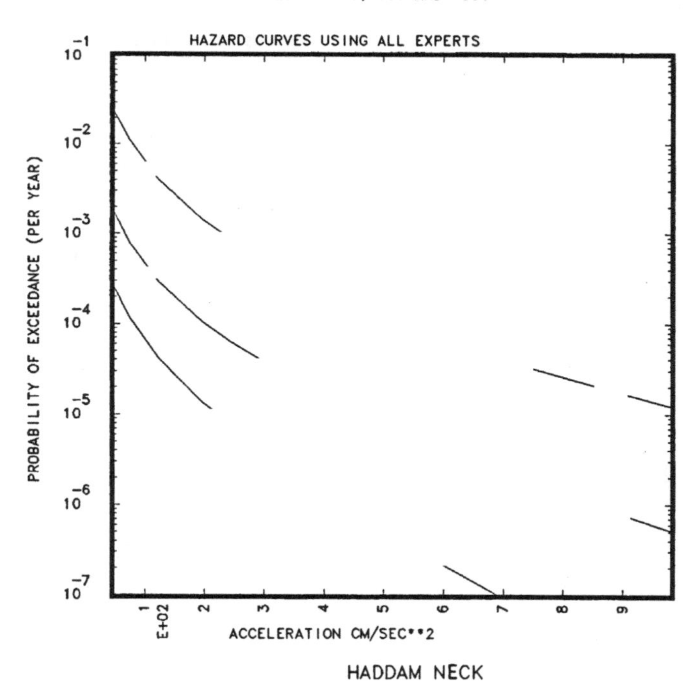

Figure 2.3.3 CPHCs for the 15th, 50th and 85th percentiles based on all S and G-Experts' input for the Haddam Neck site.

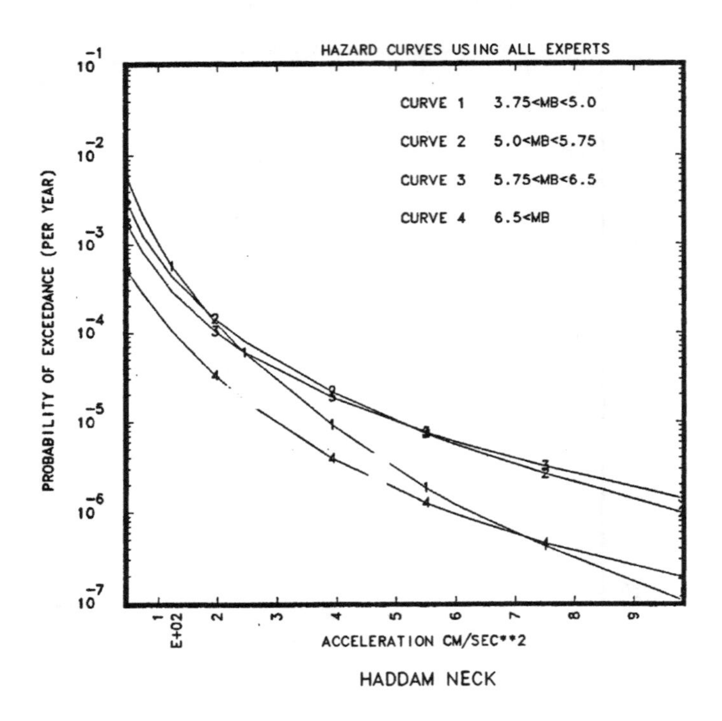

Figure 2.3.4 BEHCs which include only the contribution to the PGA hazard from earthquakes within the indicated magnitude range for the Haddam Neck site.

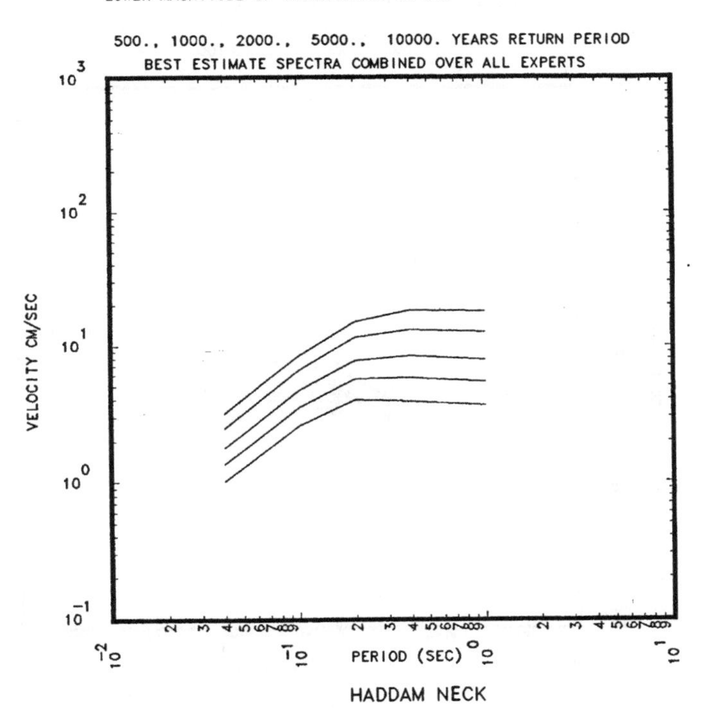

Figure 2.3.5 BEUHS for return periods of 500, 1000, 2000, 5000 and 10000 years aggregated over all S and G-Experts for the Haddam Neck site.

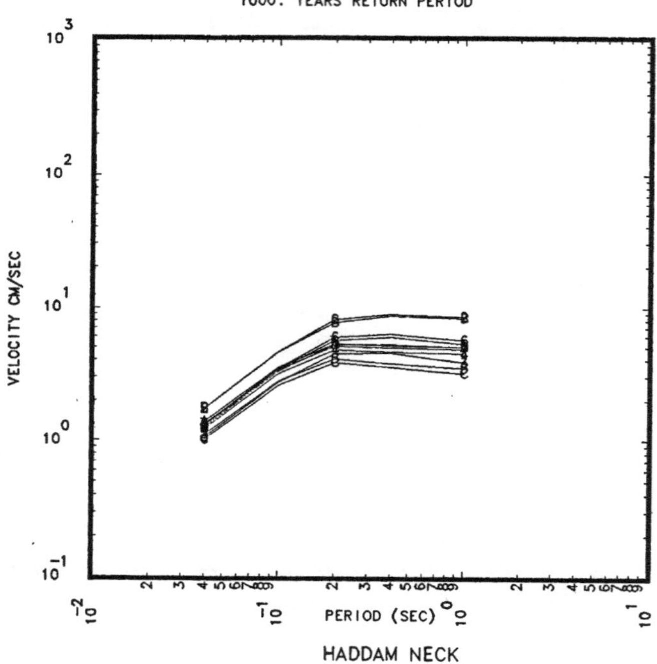

Figure 2.3.6 The 1000 year return period BEUHS per S-Expert aggregated over all G-Experts for the Haddam Neck site. Plot symbols are given in Table 2.0.

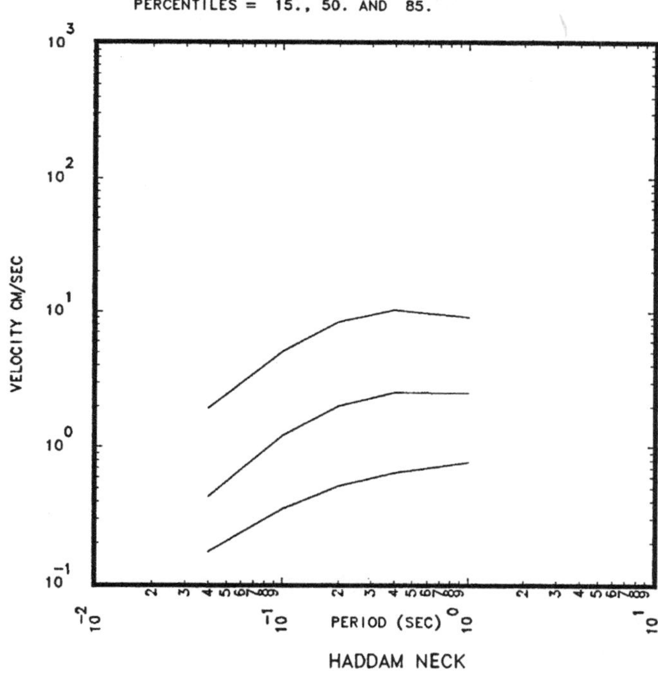

Figure 2.3.7 500 year return period CPUHS for the 15th, 50th and 85th percentiles aggregated over all S and G-Experts for the Haddam Neck site.

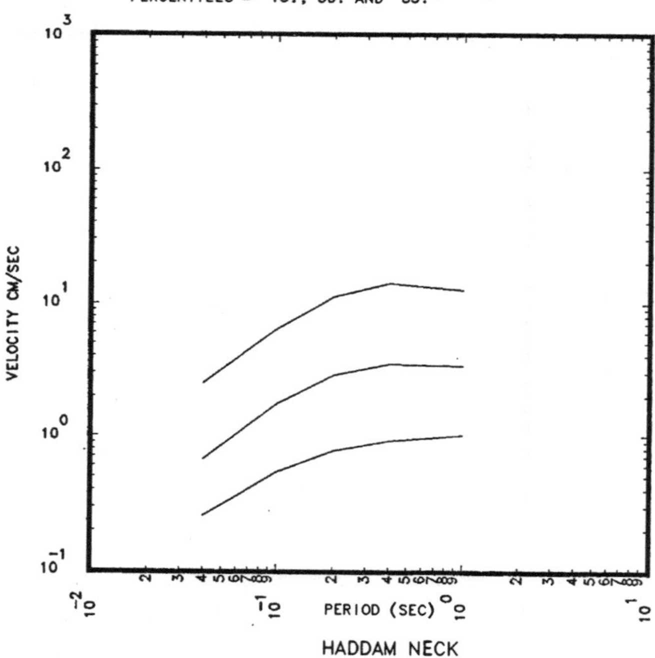

Figure 2.3.8 1000 year return period CPUHS for the 15th, 50th and 85th percentile aggregated over all S and G-Experts for the Haddam Neck site.

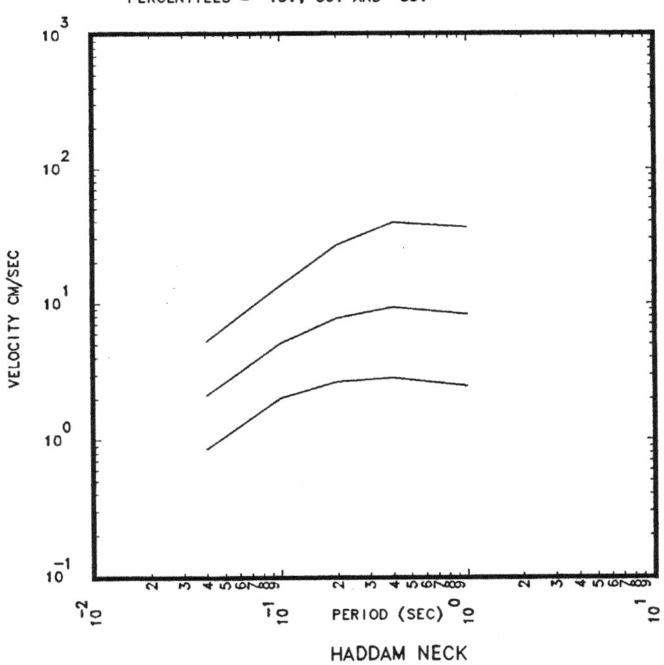

Figure 2.3.9 10000 year return period CPUHS for the 15th, 50th and 85th percentiles aggregated over all S and G-Experts for the Haddam Neck site.

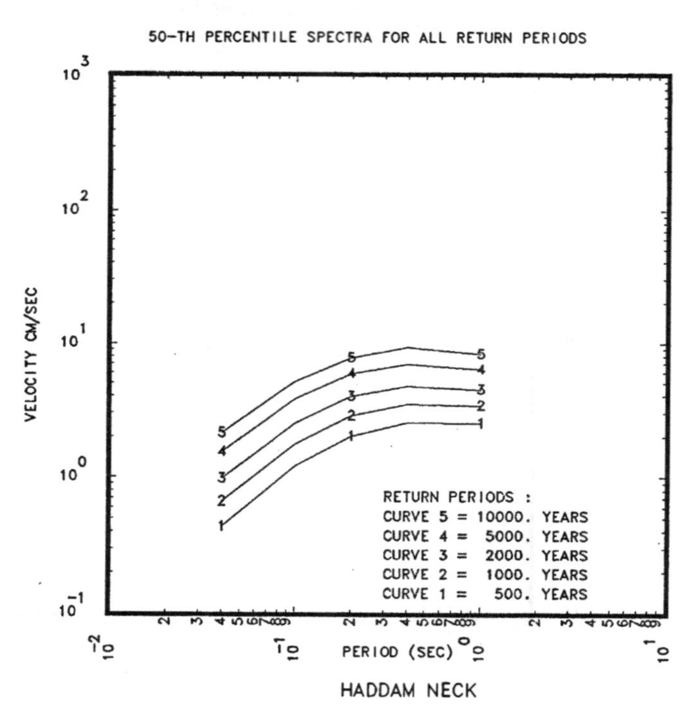

Figure 2.3.10 Comparison of the 50th percentile CPUHS for return periods of 500, 1000, 2000, 5000 and 10000 years for the Haddam Neck site.

2.4 HOPE CREEK

Hope Creek is deep soil and is represented by the symbol "4" in Fig. 1.1. It was noted in Section 2.0 that the Hope Creek site is located on the boundary between regions 1 and 2 used by our experts. G-Expert 2 did provide different BE GM models in region 1 as compared to the other regions (see Vol. I, Section 3.5). The results presented here are based on locating the site in region 1. The sensitivity to region choice is discussed in Section 3.

Table 2.4.1 and Figs. 2.4.1 to 2.4.10 give the basic results for the Hope Creek site. The AMHC is about the same as the 85th percentile CPHC. Because Hope Creek is a deep soil site then G-Expert 5's BE GM model does not dominate the BEHC and AMHC results in the same way that it did for rock sites. A typical case showing the spread between the G-Experts' BEHC for a given S-Expert's input is shown in Fig. 2.4.11. It is seen that there is very little difference between G-Experts' BEHCs. Because of this relatively small spread in the G-Experts' BEHCs the combined BEHC is relatively close to the median. The spread between G-Expert 5's BEHC and the other G-Experts' BEHC is larger for S-Expert 5 and 11. As can be seen from Table 2.4.1, for these two S-Experts zones other than the zone which contains the site are the most significant. As noted previously in Section 2.1, when this happens G-Expert 5's BEHC tends to be higher than the other G-Experts' BEHC.

It can be seen from Fig. 2.4.4 that the earthquakes in the magnitude range $5.0 \leq m_b \leq 5.75$ contribute about the same as the earthquakes in the range $5.75 \leq m_b \leq 6.5$. We also see from Fig. 2.4.4 that out to about 0.4g, including earthquakes in the range of $3.75 \leq m_b \leq 5$ would increase the BEHC by about a factor of 2.0 at 0.05g, a factor of 1.3 at 0.2g and dropping to 1. by 0.5g.

TABLE 2.4.1

MOST IMPORTANT ZONES PER S-EXPERT FOR HOPE CREEK

SITE SOIL CATEGORY DEEP-SOIL

S-XPT NUM.	HOST ZONE		ZONES CONTRIBUTING MOST SIGNIFICANTLY TO THE PGA BEHC AND % OF CONTRIBUTION AT LOW PGA(0.125G)				AT HIGH PGA(0.60G)			
1	ZONE 1	ZONE ID:	ZONE 1	ZONE 4	ZONE 20	ZONE 21	ZONE 1	ZONE 4	ZONE 3	ZONE 20
		% CON T:	66.	27.	3.	1.	95.	5.	0.	0.
2	ZONE 28	ZONE ID:	ZONE 28	ZONE 30	ZONE 32	ZONE 27	ZONE 28	COMP. ZON	ZONE 32	ZONE 30
		% CON T:	95.	2.	1.	1.	99.	0.	0.	0.
3	ZONE 8A	ZONE ID:	ZONE 8A	ZONE 5	ZONE 4	COMP. ZON	ZONE 8A	ZONE 5	COMP. ZON	ZONE 3
		% CONT.:	55.	43.	1.	1.	84.	16.	0.	0.
4	COMP. ZO	ZONE ID:	ZONE 11	ZONE 12	COMP. ZON	ZONE 16	COMP. ZON	ZONE 11	ZONE 8	ZONE 7
		% CON T:	39.	14.	13.	13.	66.	33.	1.	0.
5	ZONE 8	ZONE ID:	ZONE 1	ZONE 8	ZONE 9	ZONE 6	ZONE 1	ZONE 8	COMP. ZON	ZONE 3
		% CON T:	61.	24.	7.	7.	79.	21.	0.	0.
6	ZONE 6	ZONE ID:	ZONE 6	ZONE 7	ZONE 13	ZONE 15	ZONE 6	COMP. ZON	ZONE 7	ZONE 2
		% ONT.:	98.	1.	1.	0.	100.	0.	0.	0.
7	ZONE 29	ZONE ID:	ZONE 29	ZONE 7	ZONE 13	ZONE 14	ZONE 29	ZONE 7	ZONE 2 =	ZONE 13
		% CONT.:	84.	12.	2.	1.	95.	5.	0.	0.
10	ZONE 4B	ZONE ID:	ZONE 4B	ZONE 5	ZONE 19 =	ZONE 6	ZONE 4B	ZONE 19 =	ZONE 1	ZONE
		% CON T:	96.	2.	1.	0.	98.	2.	0.	0.
11	CZ = ZON	ZONE ID:	ZONE 5	CZ = ZONE	ZONE 3	ZONE 8	ZONE 5	CZ = ZONE	ZONE 8	ZONE 1
		% CON T:	65.	30.	2.	2.	66.	34.	0.	0.
12	ZONE 32	ZONE ID:	ZONE 32	ZONE 31	ZONE 23A	ZONE 27	ZONE 32	ZONE 17	ZONE 19	ZONE 20
		% CONT.:	99.	0.	0.	0.	100.	0.	0.	0.
13	CZ 17	ZONE ID:	CZ 17	CZ 15	ZONE 10	ZONE 9	CZ 17	CZ 15	ZONE 7	ZONE 8
		% CON T:	91.	6.	2.	0.	97.	3.	0.	0.

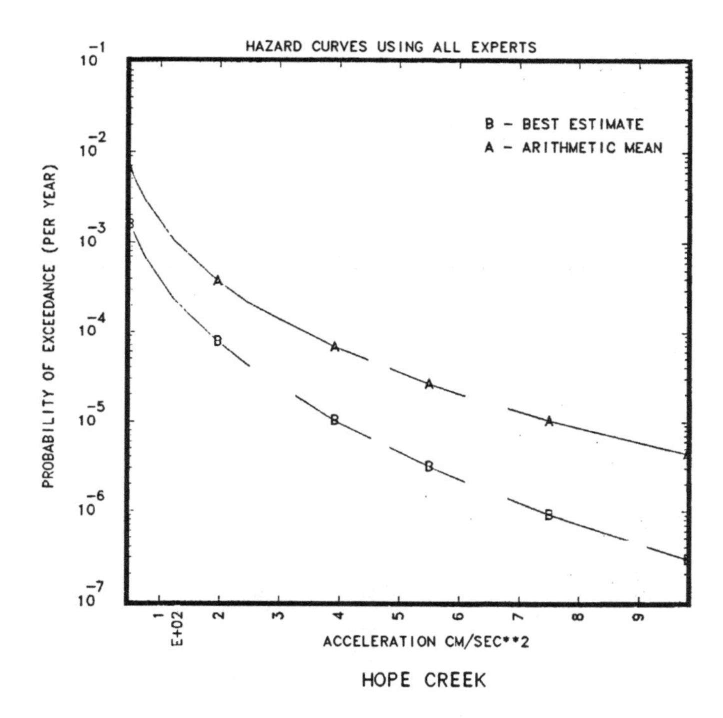

Figure 2.4.1 Comparison of the BEHC and AMHC aggregated over all S and G-Experts for the Hope Creek site.

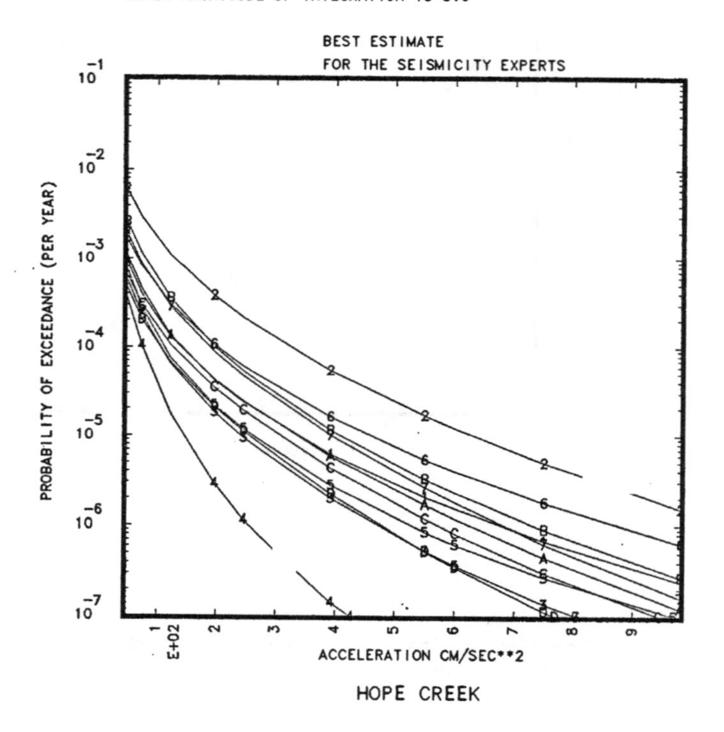

Figure 2.4.2 BEHCs per S-Expert combined over all G-Experts for the Hope Creek site. Plot symbols given in Table 2.0.

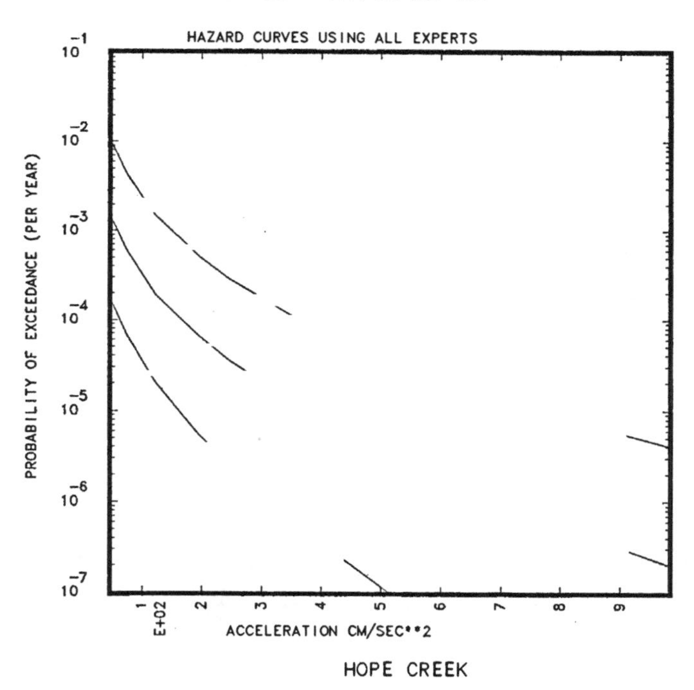

Figure 2.4.3 CPHCs for the 15th, 50th and 85th percentiles based on all S and G-Experts' input for the Hope Creek site.

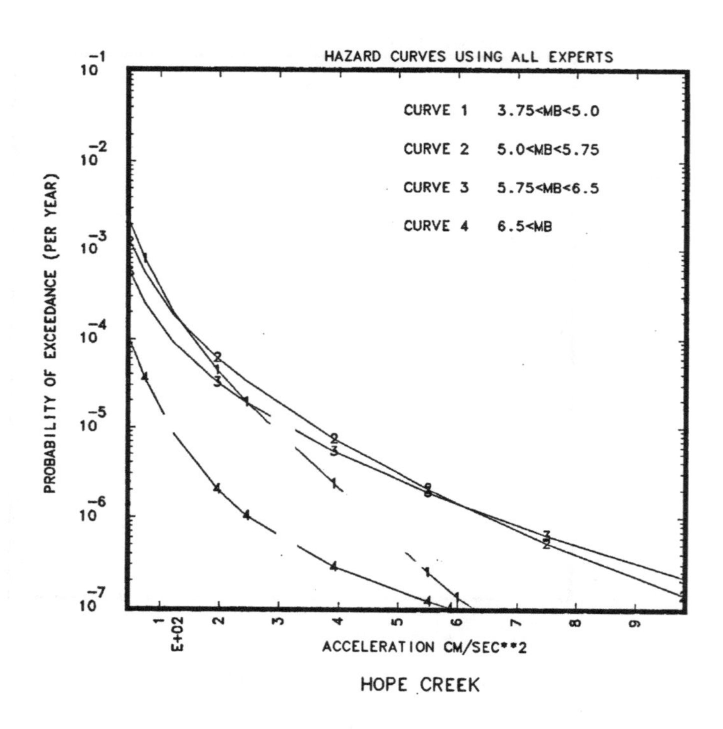

Figure 2.4.4 BEHCs which include only the contribution to the PGA hazard from earthquakes within the indicated magnitude range for the Hope Creek site.

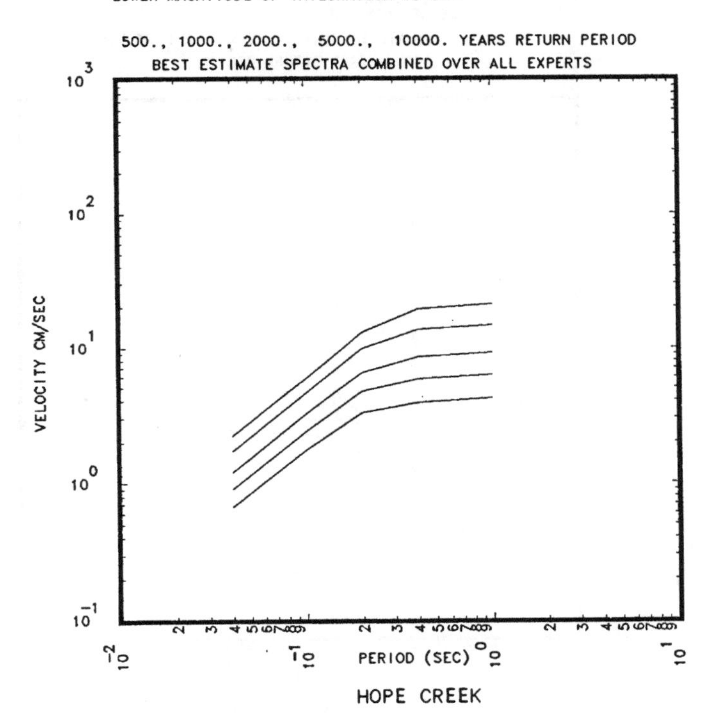

Figure 2.4.5 BEUHS for return periods of 500, 1000, 2000, 5000 and 10000 years aggregated over all S and G-Experts for the Hope Creek site.

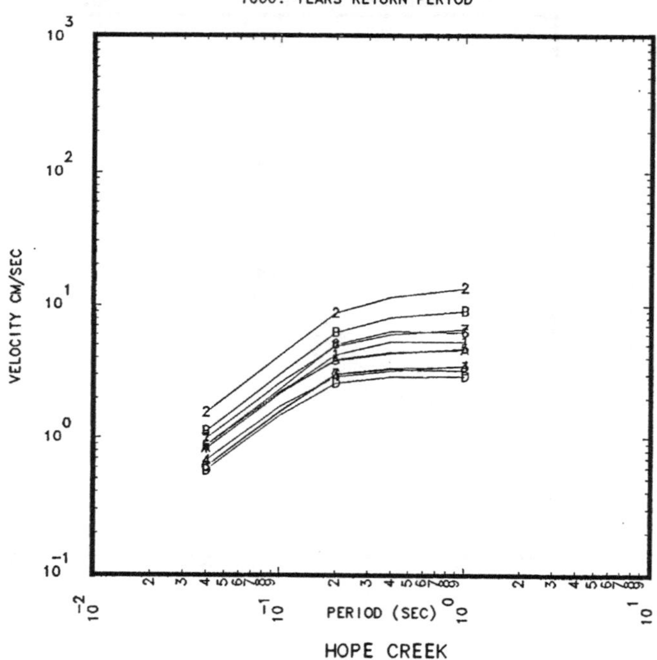

Figure 2.4.6 The 1000 year return period BEUHS per S-Expert aggregated over all G-Experts for the Hope Creek site. Plot symbols are given in Table 2.0.

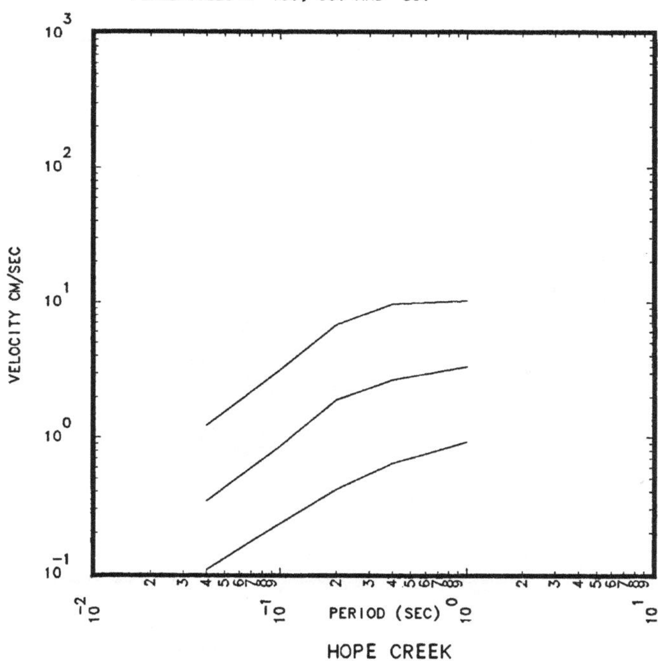

Figure 2.4.7 500 year return period CPUHS for the 15th, 50th and 85th percentiles aggregated over all S and G-Experts for the Hope Creek site.

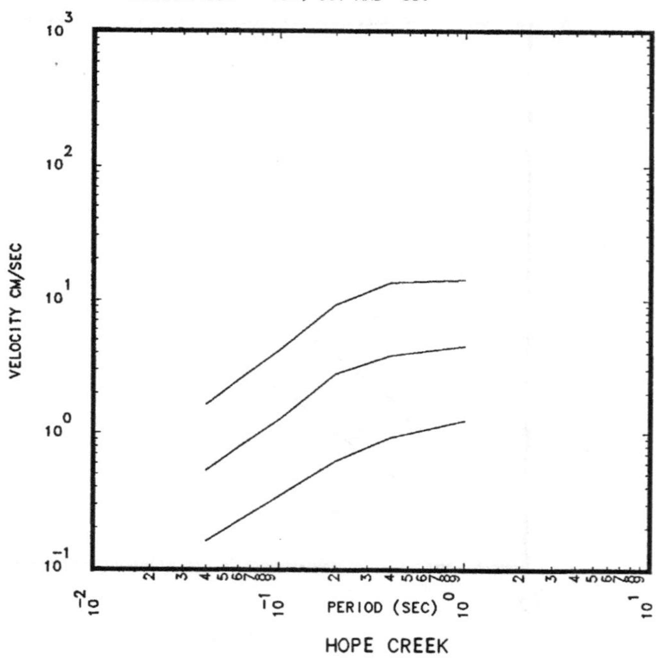

Figure 2.4.8 1000 year return period CPUHS for the 15th, 50th and 85th percentile aggregated over all S and G-Experts for the Hope Creek site.

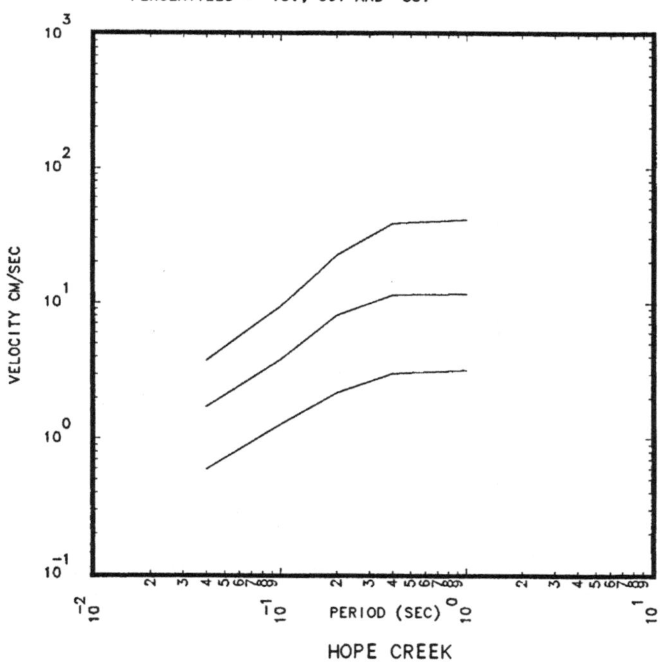

Figure 2.4.9 10000 year return period CPUHS for the 15th, 50th and 85th percentiles aggregated over all S and G-Experts for the Hope Creek site.

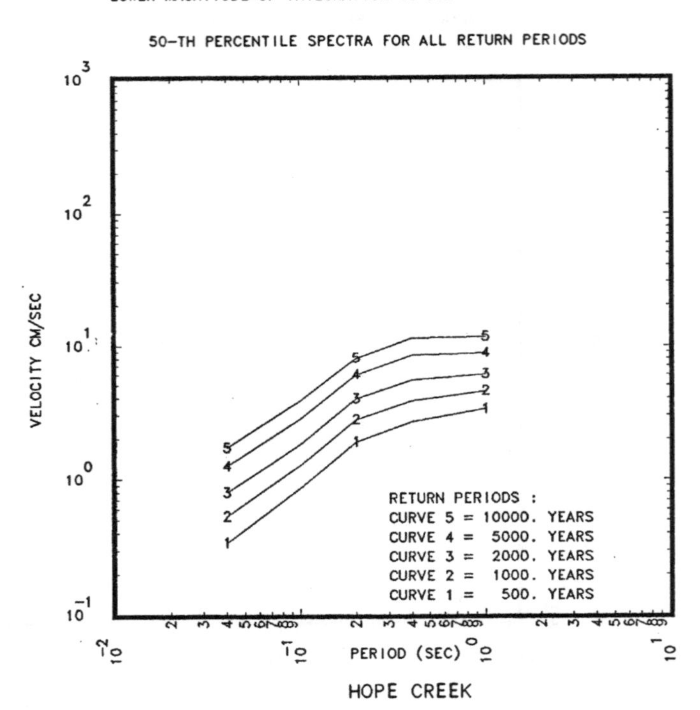

Figure 2.4.10 Comparison of the 50th percentile CPUHS for return periods of 500, 1000, 2000, 5000 and 10000 years for the Hope Creek site.

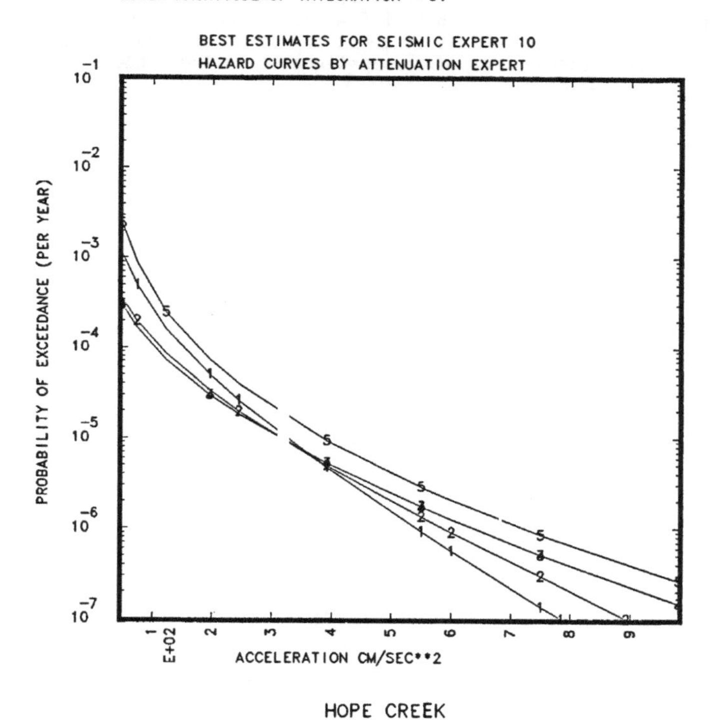

Figure 2.4.11. Comparison of the BEHCs for PGA per G-Expert for S-Expert 10's input for the Hope Creek site. The spread between the Experts' BEHCs is typical for deep soil sites in region 1.

2.5 INDIAN POINT

Indian Point is a rock site and is represented by the symbol "5" in Fig. 1.1. Table 2.5.1 and Figs. 2.5.1 to 2.5.10 give the basic results for the Indian Point site. The AMHC is about the same as the 85th percentile CPHC. Typically, the spread between the BEHCs per G-Expert for a given S-Expert is similar to that shown in Fig. 2.1.11.

We see from Fig. 2.5.4 that earthquakes in the ranges $5.0 \leq m_b < 5.75$ and $5.75 \leq m_b < 6.5$. equally contribute to the PGA hazard. We, also see that earthquakes in the range 3.75 to 5.0 would increase the PGA hazard about a factor of 2.0 at low g values to about 1.3 at about 0.3g and above about 0.5g the potential contribution of the earthquakes in the range 3.75 to 5 is small.

TABLE 2.5.1

MOST IMPORTANT ZONES PER S-EXPERT FOR INDIAN POINT

SITE SOIL CATEGORY ROCK

S-XPT NUM.	HOST ZONE		ZONES CONTRIBUTING MOST SIGNIFICANTLY TO THE PGA BEHC AND % OF CONTRIBUTION AT LOW PGA(0.125G)				AT HIGH PGA(0.60G)			
1	ZONE 4	ZONE ID:	ZONE 4	ZONE 20	ZONE 22	ZONE 21	ZONE 4	ZONE 22	ZONE 20	ZONE 1
		% CON T:	52.	21.	15.	7.	72.	16.	8.	3.
2	ZONE 31	ZONE ID:	ZONE 31	ZONE 28	ZONE 32	COMP. ZON	ZONE 31	ZONE 28	ZONE 32	COMP. Z
		% CONT.:	56.	24.	17.	2.	82.	9.	9.	1.
3	ZONE 4	ZONE ID:	ZONE 4	ZONE 5	ZONE 2	ZONE 3	ZONE 4	ZONE 5	ZONE 2	ZONE 8A
		% CONT.:	59.	31.	5.	3.	76.	24.	1.	0.
4	ZONE 11	ZONE ID:	ZONE 1	ZONE 16	ZONE 18	ZONE 19	ZONE 11	ZONE 16	ZONE 12	ZONE 18
		% CONT.:	37.	24.	13.	7.	91.	6.	1.	1.
5	ZONE 1	ZONE ID:	ZONE 1	ZONE 6	ZONE 3	ZONE 4	ZONE 1	ZONE 3	ZONE 4	COMP. Z
		% CONT.:	53.	41.	3.	1.	9.	1.	0.	0.
6	ZONE 6	ZONE ID:	ZONE 6	ZONE 7	ZONE 3	ZONE 5	ZONE 6	ZONE 7	ZONE 5	ZONE 3
		% CONT.:	77.	13.	6.	3.	94.	5.	1.	0.
7	ZONE 24	ZONE ID:	ZONE 13	ZONE 14	ZONE 17	ZONE 24	ZONE 13	ZONE 14	ZONE 24	ZONE 41
		% CONT.:	38.	19.	10.	7.	62.	16.	11.	4.
10	ZONE 5	ZONE ID:	ZONE 5	ZONE 6	ZONE 4B	ZONE 4A	ZONE 5	ZONE 4A	ZONE 4B	ZONE 19
		% CONT.:	51.	9.	8.	8.	77.	12.	5.	3.
11	ZONE 5	ZONE ID:	ZONE 5	ZONE 3	ZONE 4	CZ = ZONE	ZONE 5	ZONE 4	ZONE 3	CZ = ZO
		% CON T:	68.	17.	8.	4.	94.	3.	2.	1.
12	ZONE 32	ZONE ID:	ZONE 32	ZONE 31	ZONE 34	ZONE 27	ZONE 32	ZONE 27	ZONE 34	ZONE 31
		% CONT.:	73.	13.	7.	4.	98.	1.	0.	0.
13	ZONE 10	ZONE ID:	ZONE 10	CZ 15	ZONE 12	ZONE 11	ZONE 10	CZ 15	ZONE 12	CZ 17
		% CONT.:	86.	5.	5.	3.	98.	2.	0.	0.

E U.S. SEISMIC HAZARD CHARACTERIZATION
LOWER MAGNITUDE OF INTEGRATION IS 5.0

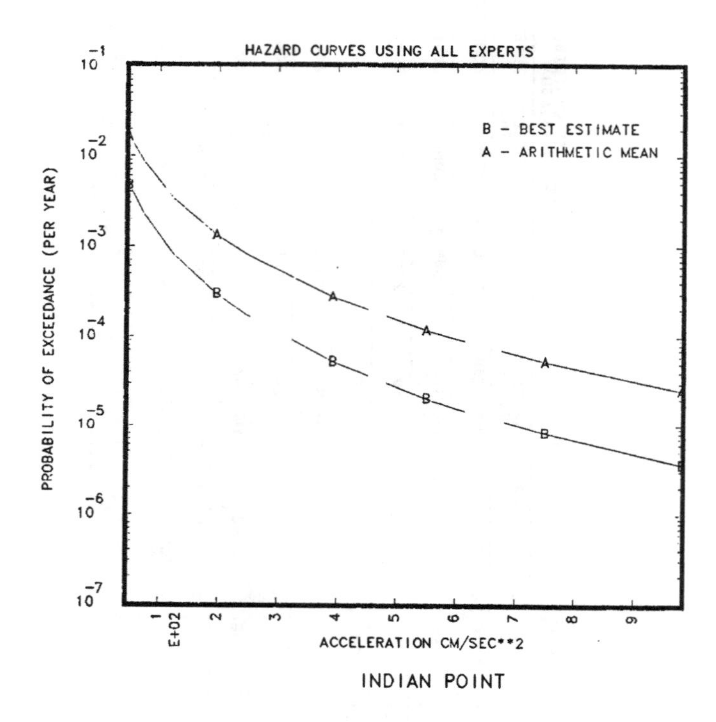

Figure 2.5.1 Comparison of the BEHC and AMHC aggregated over all S and G-Experts for the Indian Point site.

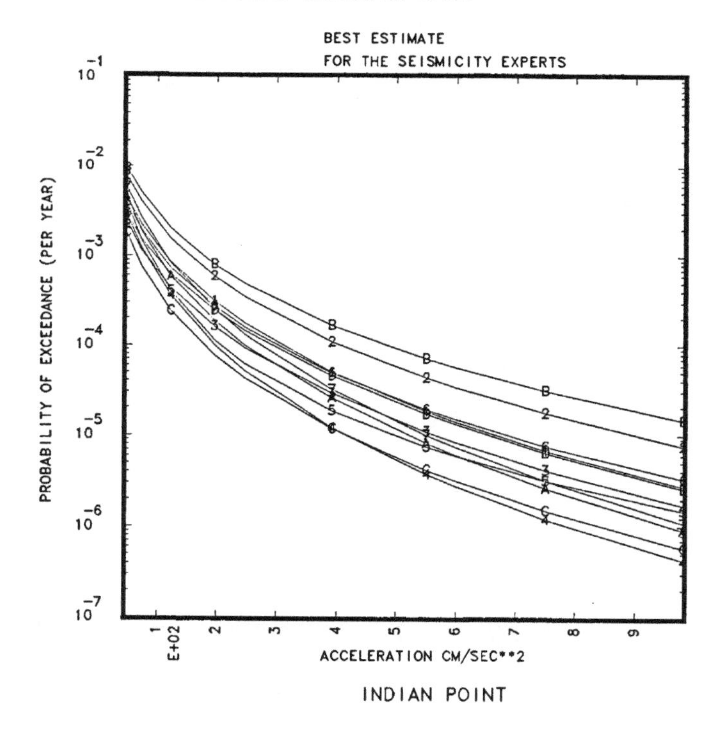

Figure 2.5.2 BEHCs per S-Expert combined over all G-Experts for the Indian Point site. Plot symbols given in Table 2.0.

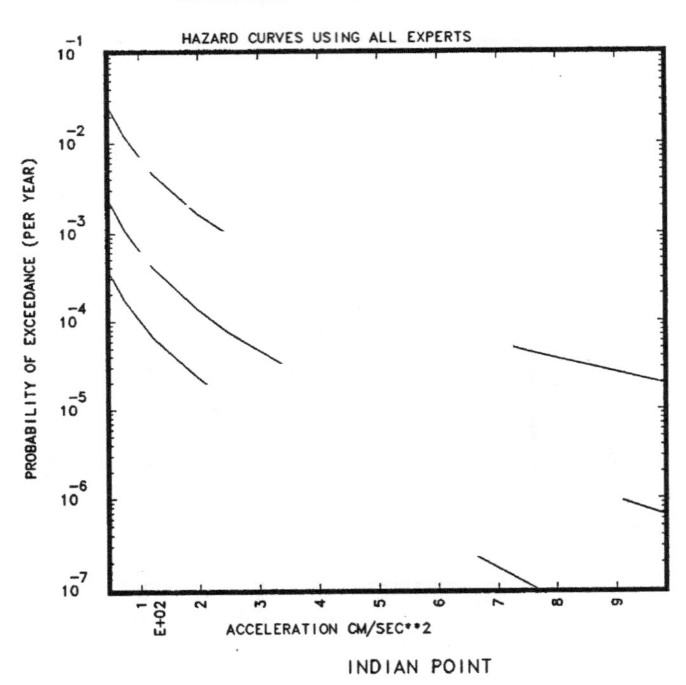

Figure 2.5.3 CPHCs for the 15th, 50th and 85th percentiles based on all S and G-Experts' input for the Indian Point site.

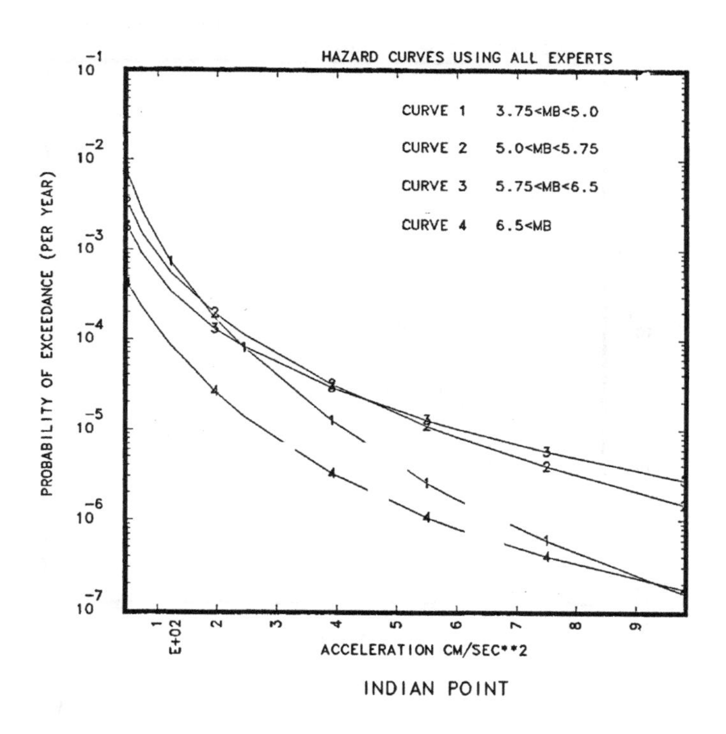

Figure 2.5.4 BEHCs which include only the contribution to the PGA hazard from earthquakes within the indicated magnitude range for the Indian Point site.

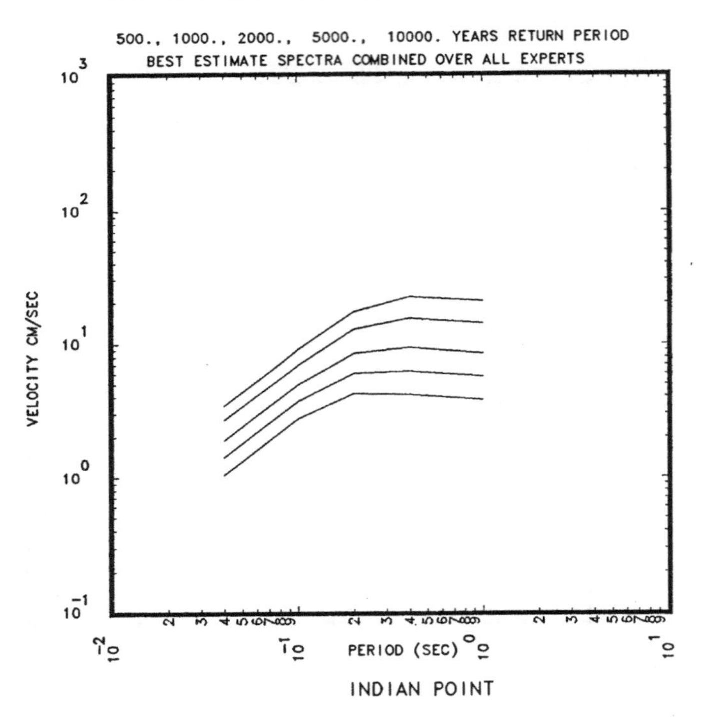

Figure 2.5.5 BEUHS for return periods of 500, 1000, 2000, 5000 and 10000 years aggregated over all S and G-Experts for the Indian Point site.

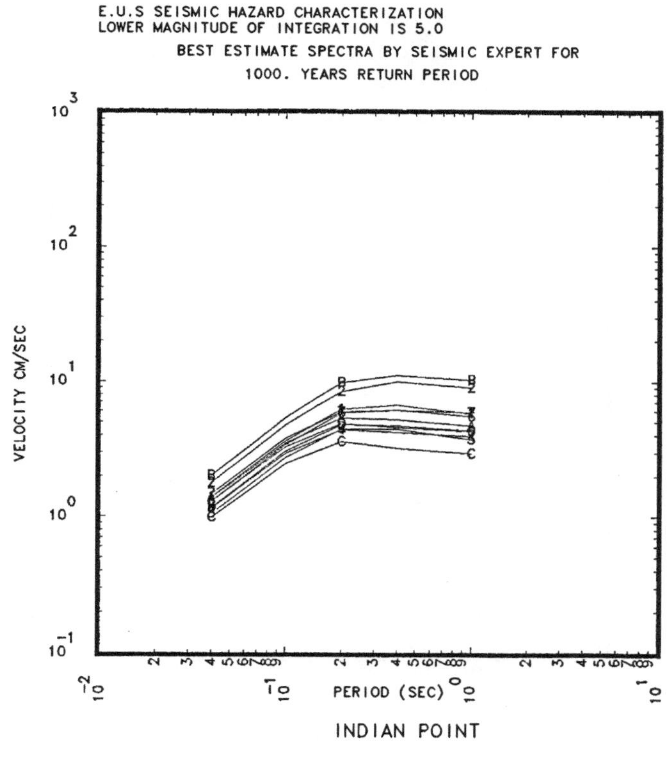

E.U.S SEISMIC HAZARD CHARACTERIZATION
LOWER MAGNITUDE OF INTEGRATION IS 5.0

BEST ESTIMATE SPECTRA BY SEISMIC EXPERT FOR
1000. YEARS RETURN PERIOD

VELOCITY CM/SEC

PERIOD (SEC)

INDIAN POINT

Figure 2.5.6 The 1000 year return period BEUHS per S-Expert aggregated over all G-Experts for the Indian Point site. Plot symbols are given in Table 2.0.

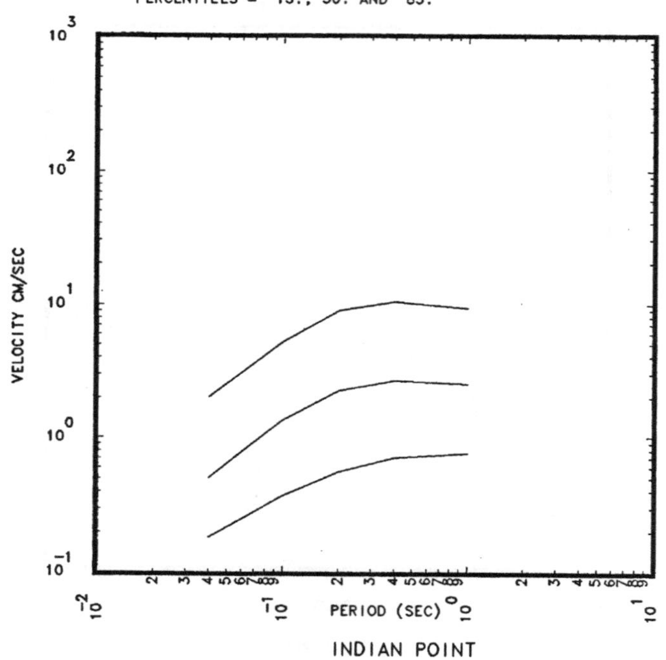

Figure 2.5.7 500 year return period CPUHS for the 15th, 50th and 85th percentiles aggregated over all S and G-Experts for the Indian Point site.

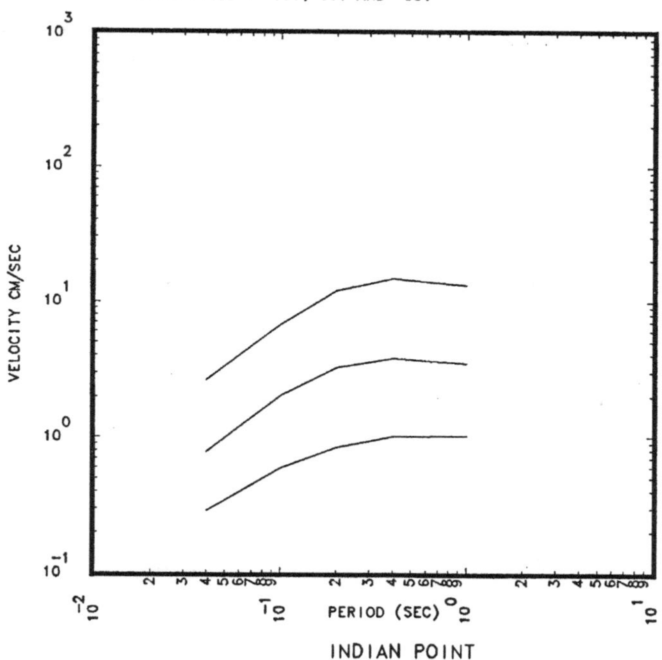

Figure 2.5.8 1000 year return period CPUHS for the 15th, 50th and 85th percentile aggregated over all S and G-Experts for the Indian Point site.

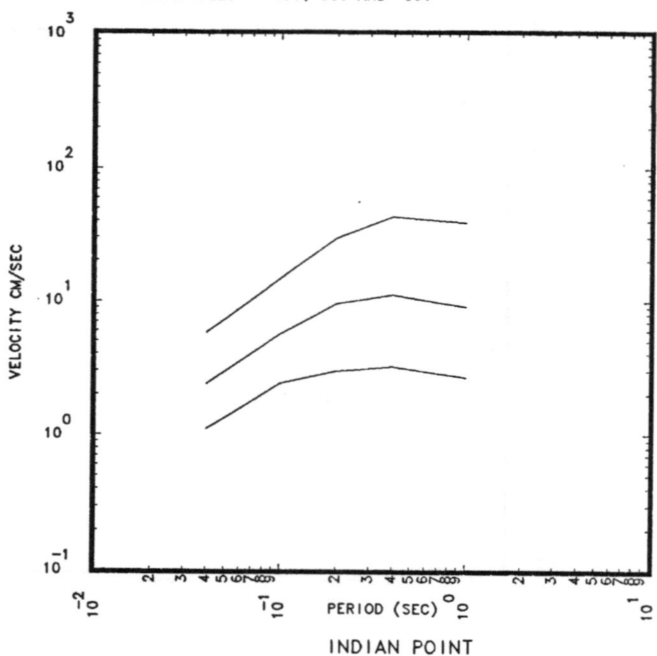

Figure 2.5.9 10000 year return period CPUHS for the 15th, 50th and 85th percentiles aggregated over all S and G-Experts for the Indian Point site.

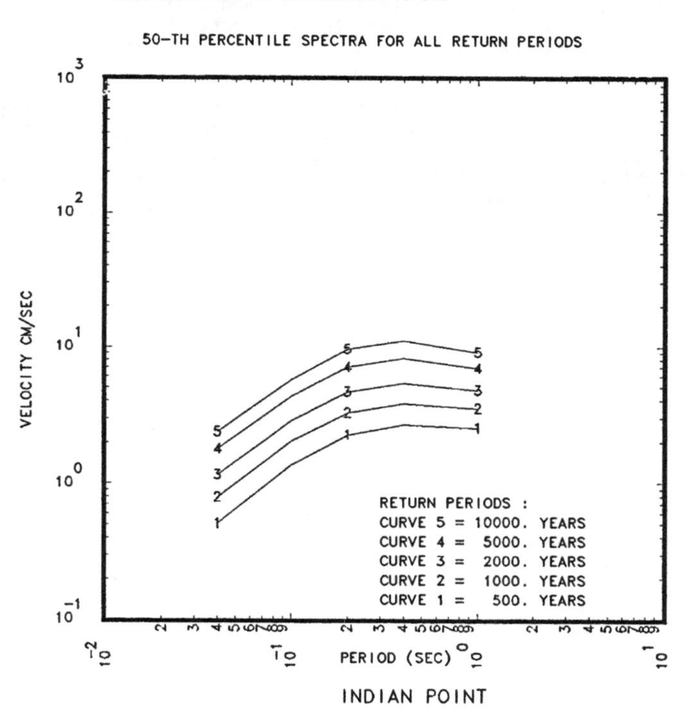

Figure 2.5.10 Comparison of the 50th percentile CPUHS for return periods of 500, 1000, 2000, 5000 and 10000 years for the Indian Point site.

2.6 LIMERICK

Limerick is a rock site and is represented by the symbol "6" in Fig. 1.1. Table 2.6.1 and Figs. 2.6.1 to 2.6.10 give the basic results for the Limerick site. Limerick was one of the ten test sites analyzed earlier for which the results are documented in Bernreuter et. al., (1985), documented in UCID-20421, Vols. 1 and 2, "Seismic Hazard Characterization of the Eastern United States". As indicated in Vol. I there is very little difference between the PGA AMHC given in Fig. 2.6.1 and our previous results reported in Bernreuter et al., 1985, provided that the lower bound of magnitude used for integration is the same. However the UHCPS are significantly different at longer periods. The spread between the G-Experts' BEHCs is similar to the spread shown in Fig. 2.1.11.

The AMHC is about the same as the 85th percentile CPHC. We see from Fig. 2.6.2 that there is a wide diversity among the S-Experts. From Fig. 2.6.4 we see that earthquakes in the range of 3.75 to 5.0 would contribute significantly to the PGA hazard. Thus including the earthquakes between magnitude 3.75 and 5 would multiply the final seismic hazard by about a factor of 2.0 at 0.05g, a factor of 1.3 at 0.2g, and nearly 1.0 at 0.5g.

TABLE 2.6.1

MOST IMPORTANT ZONES PER S-EXPERT FOR LIMERICK

SITE SOIL CATEGORY ROCK

S-XPT NUM.	HOST ZONE		ZONES CONTRIBUTING MOST SIGNIFICANTLY TO THE PGA BEHC AND % OF CONTRIBUTION AT LOW PGA(0.125G)				AT HIGH PGA(0.60G)			
1	ZONE 4	ZONE ID:	ZONE 4	ZONE 20	ZONE 1	ZONE 21	ZONE 4	ZONE 1	ZONE 20	ZONE 2
		% CONT.:	79.	10.	5.	3.	95.	3.	1.	0.
2	ZONE 28	ZONE ID:	ZONE 28	ZONE 32	ZONE 31	ZONE 30	ZONE 28	ZONE 32	COMP. ZON	ZONE 3
		% CON .T:	84.	6.	4.	3.	96.	2.	1.	1.
3	ZONE 5	ZONE ID:	ZONE 5	ZONE 4	ZONE 8A	ZONE 2	ZONE 5	ZONE 8A	ZONE 4	ZONE 2
		% CONT.:	88.	4.	3.	2.	99.	1.	0.	0.
4	ZONE 11	ZONE ID:	ZONE 11	ZONE 12	ZONE 16	ZONE 8	ZONE 11	ZONE 12	ZONE 16	ZONE
		% CONT.:	44.	22.	16.	5.	85.	13.	1.	1.
5	ZONE 1	ZONE ID:	ZONE 6	ZONE 1	ZONE 9	ZONE 8	ZONE 1	ZONE 9	ZONE 6	ZONE
		% CON T.:	52.	40.	4.	1.	99.	0.	0.	0.
6	ZONE 6	ZONE ID:	ZONE 6	ZONE 7	ZONE 13	ZONE 3	ZONE 6	ZONE 7	COMP. ZON	ZONE 1
		% CON .T:	86.	7.	2.	2.	99.	1.	0.	0.
7	ZONE 7	ZONE ID:	ZONE 7	ZONE 13	ZONE 29	ZONE 14	ZONE 7	ZONE 29	ZONE 13	ZONE 1
		% CON T.:	62.	13.	12.	4.	93.	4.	3.	0.
10	ZONE 4B	ZONE ID:	ZONE 5	ZONE 4B	ZONE 19 =	ZONE 6	ZONE 5	ZONE 8	ZONE 19 =	ZONE
		% CON T.:	57.	30.	4.	3.	54.	44.	2.	0.
11	ZONE 5	ZONE ID:	ZONE 5	ZONE 3	CZ = ZONE	ZONE 4	ZONE 5	ZONE 4	ZONE 3	CZ = Z
		% CON .T:	83.	8.	3.	3.	99.	0.	0.	0.
12	ZONE 32	ZONE ID:	ZONE 32	ZONE 27	ZONE 31	ZONE 34	ZONE 32	ZONE 27	ZONE 34	ZONE 1
		% CONT.:	82.	7.	7.	3.	98.	2.	0.	0.
13	CZ 15	ZONE ID:	CZ 15	ZONE 10	CZ 17	ZONE 12	CZ 15	ZONE 10	CZ 17	ZONE 1
		% CON .T:	48.	24.	15.	7.	91.	5.	3.	0.

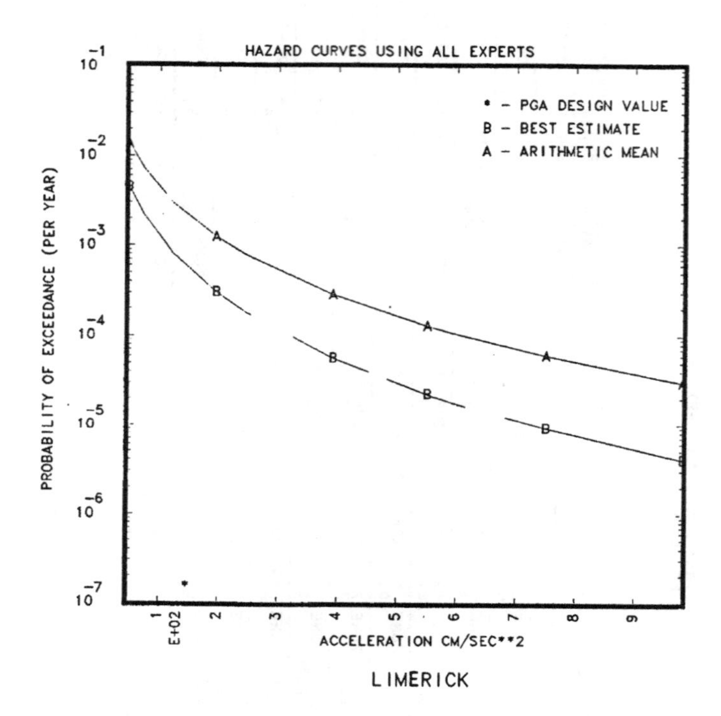

Figure 2.6.1 Comparison of the BEHC and AMHC aggregated over all S and G-Experts for the Limerick site.

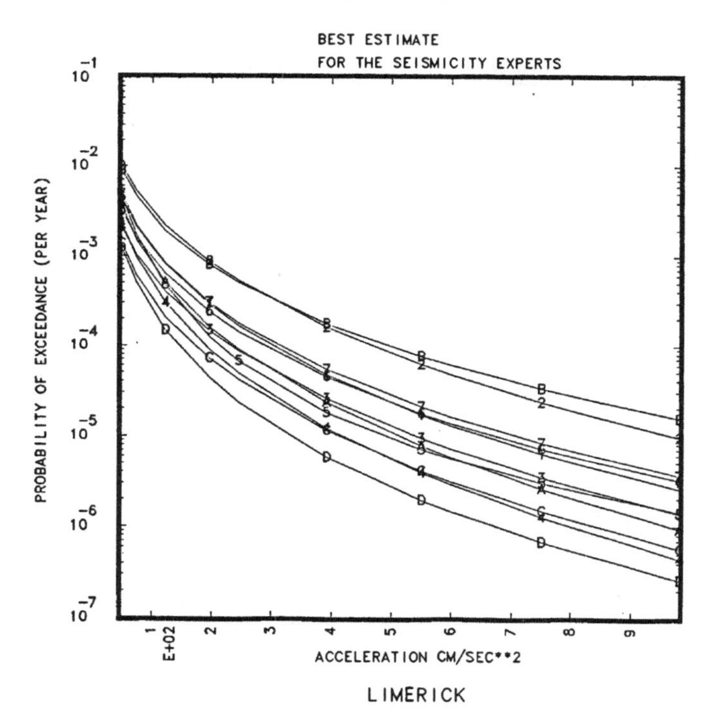

Figure 2.6.2 BEHCs per S-Expert combined over all G-Experts for the Limerick site. Plot symbols given in Table 2.0.

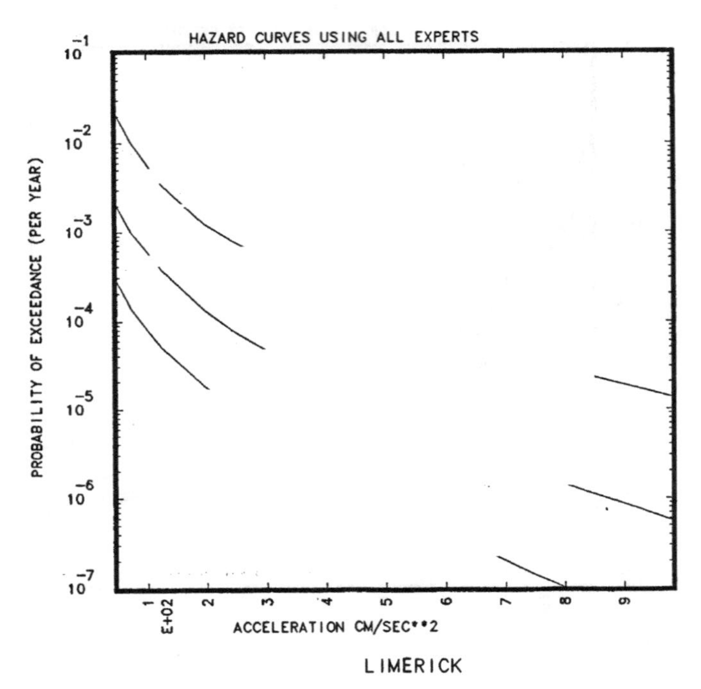

Figure 2.6.3 CPHCs for the 15th, 50th and 85th percentiles based on all S and G-Experts' input for the Limerick site.

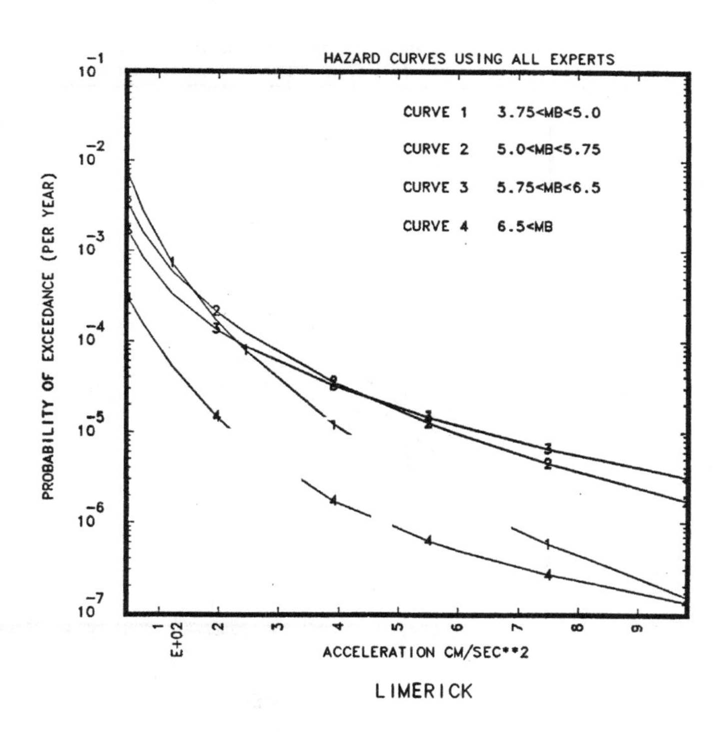

Figure 2.6.4 BEHCs which include only the contribution to the PGA hazard from earthquakes within the indicated magnitude range for the Limerick site.

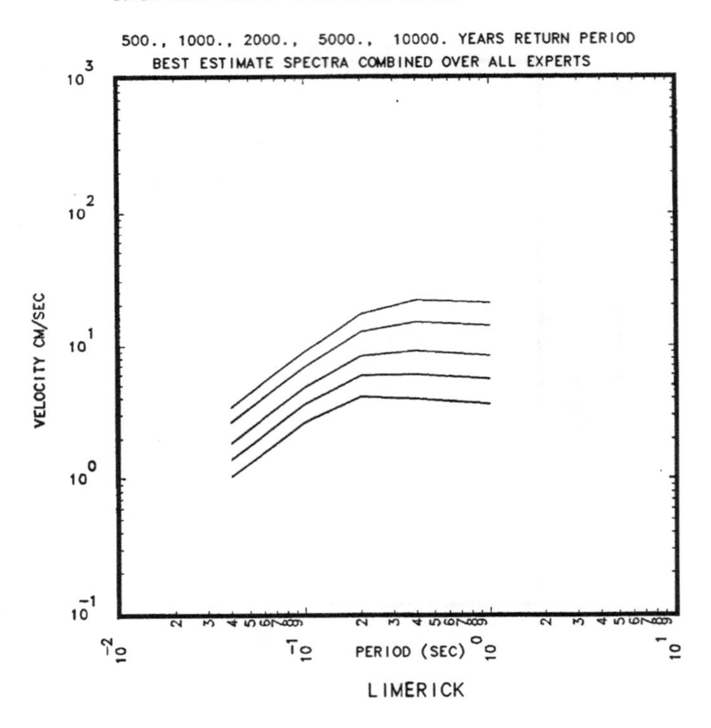

Figure 2.6.5 BEUHS for return periods of 500, 1000, 2000, 5000 and 10000 years aggregated over all S and G-Experts for the Limerick site.

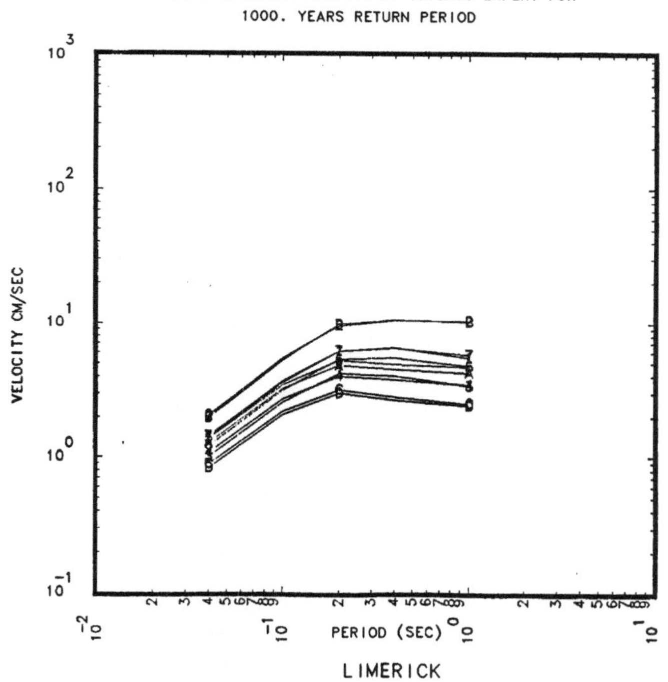

Figure 2.6.6 The 1000 year return period BEUHS per S-Expert aggregated over all G-Experts for the Limerick site. Plot symbols are given in Table 2.0.

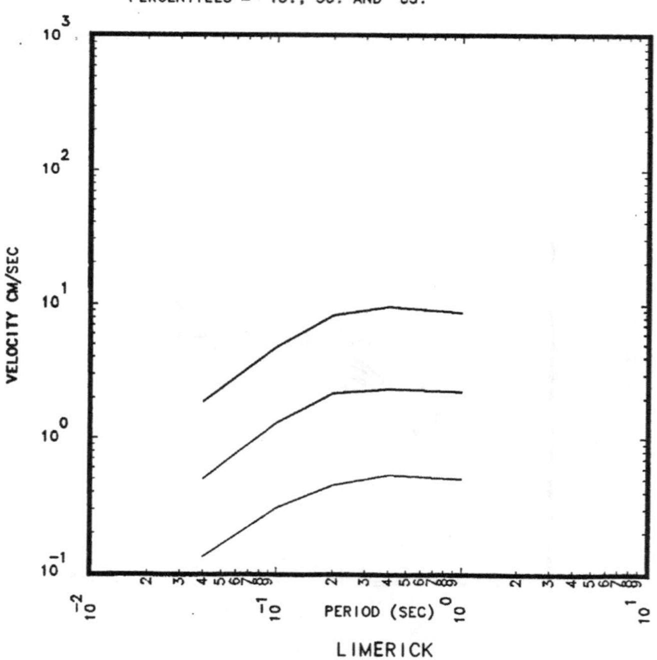

Figure 2.6.7 500 year return period CPUHS for the 15th, 50th and 85th percentiles aggregated over all S and G-Experts for the Limerick site.

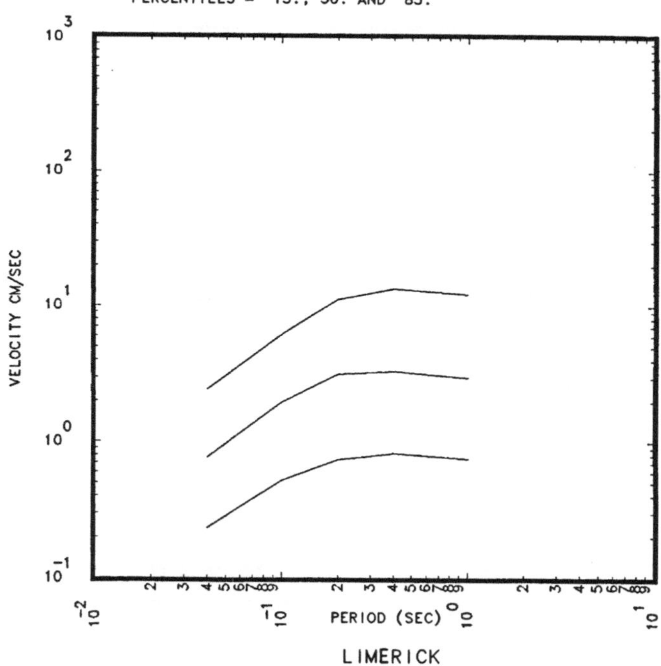

Figure 2.6.8 1000 year return period CPUHS for the 15th, 50th and 85th percentile aggregated over all S and G-Experts for the Limerick site.

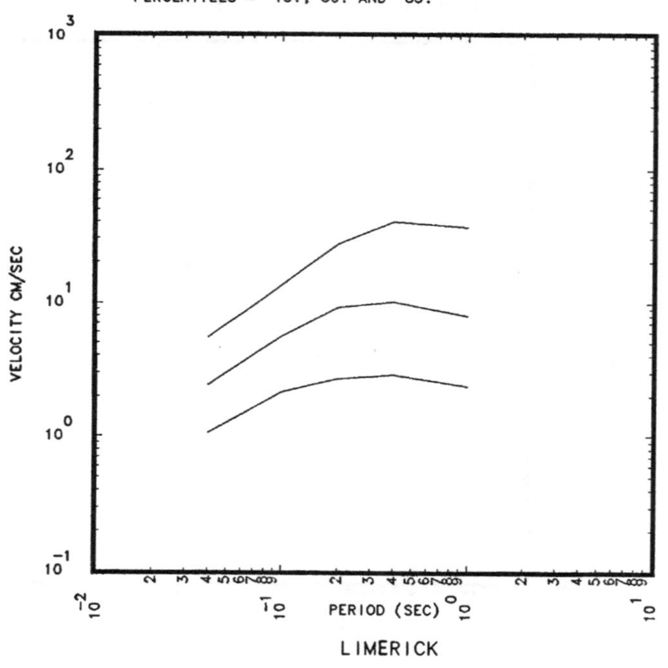

Figure 2.6.9 10000 year return period CPUHS for the 15th, 50th and 85th percentiles aggregated over all S and G-Experts for the Limerick site.

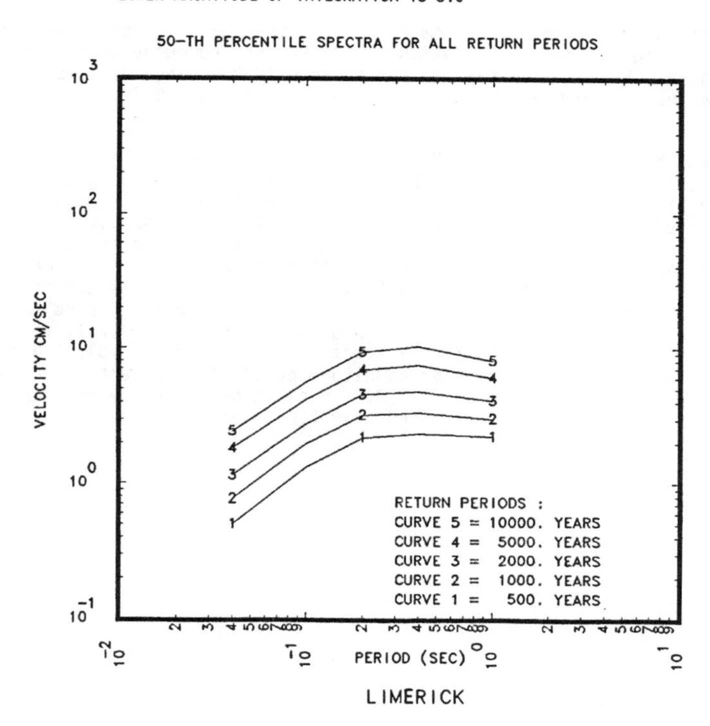

Figure 2.6.10 Comparison of the 50th percentile CPUHS for return periods of 500, 1000, 2000, 5000 and 10000 years for the Limerick site.

2.7 MAINE YANKEE

Maine Yankee is a rock site and is indicated by the symbol "7" in Fig. 1.1. Table 2.7.1 and Figs. 2.71 to 2.7.10 give the basic results of the analysis. The Maine Yankee site was one of the ten test sites analyzed in Bernreuter et al. (1985). As discussed in Vol. 1, Section 4 there is little difference between the AMHC and the CPHCs plotted in Figs. 2.7.1 and 2.73 and our previous results, see Bernreuter et. al., (1985), for a lower bound magnitude of 5. However, the CPUHS are significantly different from our previous results.

It is interesting to observe from Fig. 2.7.4 that earthquakes larger than m_b = 6.5 make a significant contribution to the PGA hazard curve at the Maine Yankee site. We also see that if earthquakes in the range of 3.75 to 5 were included the hazard would only be affected out to about 0.2g.

The spread between G-Experts' BEHCs is relatively similar to that shown in Fig. 2.1.11 except for S-Experts 1,2,7,11 and 13 where it is somewhat larger (about a factor of 2 on probability of exceedance), but still much less than the extremely large difference shown in Fig. 2.1.12. Generally, the increase in relative spread between G-Expert 5's BEHC and the others for S-Experts 1,2,7,11 and 13 occurs because some distant zone or zones with large earthquakes combined with the low attenuation of G-Expert 5's ground motion model, contribute significantly to the hazard. For example, Maine Yankee is in S-Expert 11's zone 1 however, as can be seen from Table 2.7.1, zone 3 contributes most to the BEHC for S-Expert 11. Referring to Vol. 1 Appendix B, we find that the BE for the upper magnitude cutoff in S-Expert 11's zone 1 is m_b = 5.8 whereas for zone 3 it is 7.0. In addition the slope for the b-value is much different, 1.0 in zone 1 and only 0.6 in zone 3.

TABLE 2.7.1

MOST IMPORTANT ZONES PER S-EXPERT
FOR MAINE YANKEE

SITE SOIL CATEGORY ROCK

S-XPT NUM.	HOST ZONE		ZONES CONTRIBUTING MOST SIGNIFICANTLY TO THE PGA AT LOW PGA(0.125G)				EBC AND % OF CONTRIBUTION AT HIGH PGA(0.60G)			
1	ZONE 22	ZONE ID:	ZONE 22	ZONE 21	ZONE 20	ZONE 4	ZONE 22	ZONE 21	ZONE 20	ZONE 1
		% CONT.:	54.	30.	15.	0.	73.	24.	3.	0.
2	ZONE 31	ZONE ID:	ZONE 31	ZONE 32	COMP. ZON	ZONE 28	ZONE 31	ZONE 32	COMP. ZON	ZONE 27
		% CONT.:	53.	46.	1.	0.	53.	46.	0.	0.
3	ZONE 4	ZONE ID:	ZONE 4	ZONE 3	ZONE 2	COMP. ZON	ZONE 4	ZONE 3	ZONE 2	COMP. ZON
		% CONT.:	73.	21.	6.	0.	94.	6.	0.	0.
4	ZONE 20	ZONE ID:	ZONE 20	ZONE 18	ZONE 19	ZONE 16	ZONE 20	ZONE 18	ZONE 19	ZONE 16
		% CONT.:	58.	28.	10.	5.	84.	14.	1.	0.
5	ZONE 1	ZONE ID:	ZONE 1	ZONE 3	ZONE 6	ZONE 4	ZONE 1	ZONE 3	ZONE 4	COMP. ZON
		% CONT.:	74.	16.	8.	1.	92.	8.	0.	0.
6	ZONE 5	ZONE ID:	ZONE 5	ZONE 3	ZONE 7	ZONE 4	ZONE 5	ZONE 3	ZONE 7	ZONE 4
		% CONT.:	58.	32.	6.	2.	84.	15.	1.	0.
7	ZONE 24	ZONE ID:	ZONE 19	ZONE 26	ZONE 24	ZONE 17	ZONE 19	ZONE 24	ZONE 26	ZONE 36
		% CON J:	38.	25.	16.	8.	47.	29.	20.	2.
10	ZONE 1	ZONE ID:	ZONE 23	ZONE 1	ZONE 21	ZONE 8	ZONE 23	ZONE 1	ZONE 21	ZONE 8
		% CONT.:	37.	23.	13.	13.	52.	40.	6.	2.
11	ZONE 1	ZONE ID:	ZONE 3	ZONE 1	CZ = ZONE	ZONE 2	ZONE 3	ZONE 1	CZ = ZONE	ZONE 5
		% CON J:	68.	29.	1.	1.	68.	31.	0.	0.
12	ZONE 32	ZONE ID:	ZONE 32	ZONE 34	ZONE 31	ZONE 33	ZONE 32	ZONE 34	ZONE 38	ZONE 33
		% CONT.:	41.	27.	13.	11.	74.	16.	6.	3.
13	CZ 15	ZONE ID:	ZONE 10	ZONE 12	CZ 15	ZONE 11	ZONE 10	ZONE 12	CZ 15	ZONE 11
		% CONT.:	44.	41.	12.	3.	62.	22.	15.	0.

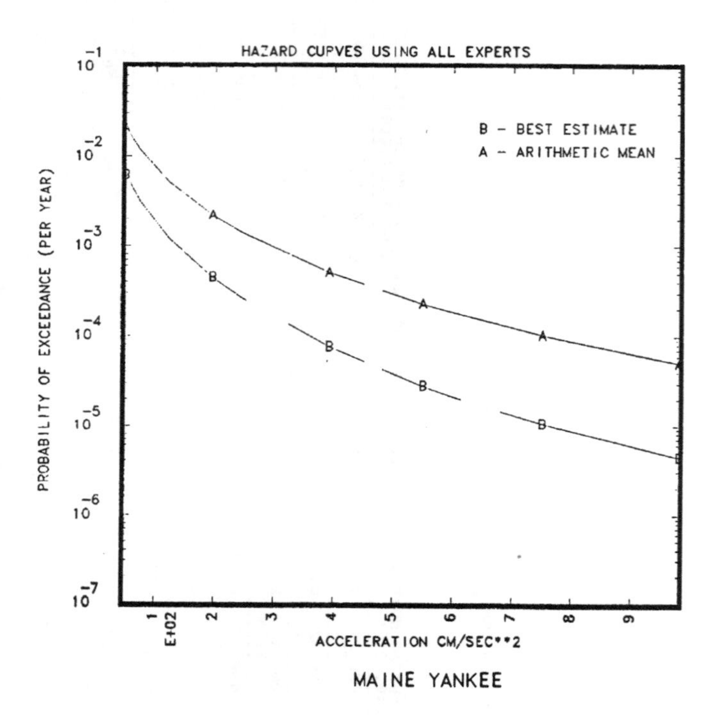

Figure 2.7.1 Comparison of the BEHC and AMHC aggregated over all S and G-Experts for the Maine Yankee site.

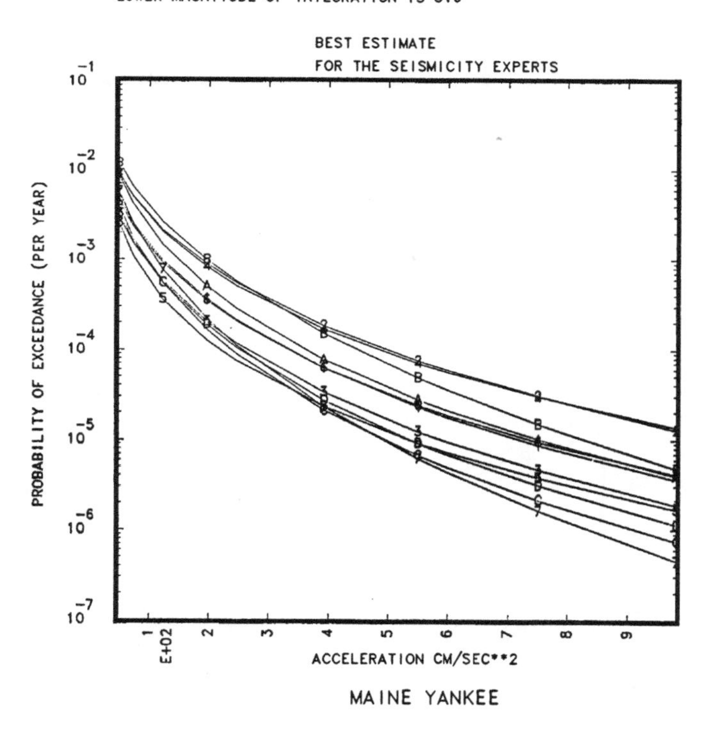

Figure 2.7.2 BEHCs per S-Expert combined over all G-Experts for the Maine Yankee site. Plot symbols given in Table 2.0.

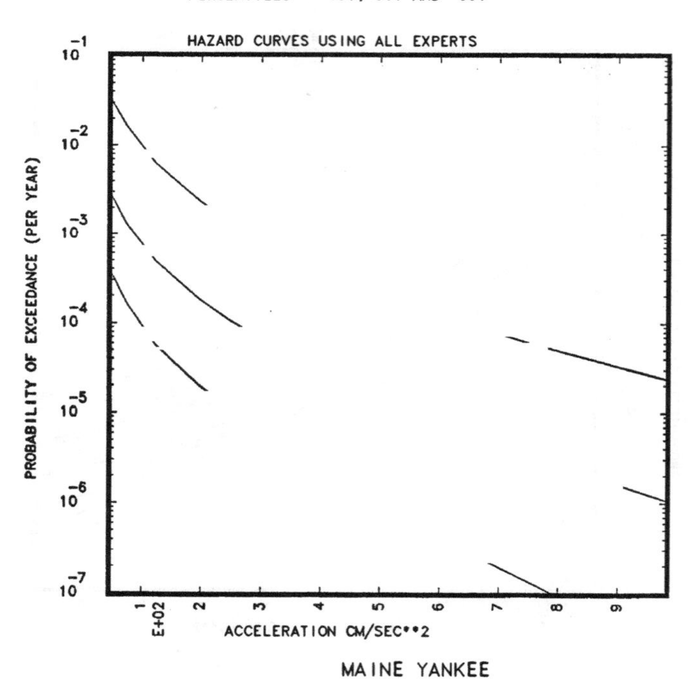

Figure 2.7.3 CPHCs for the 15th, 50th and 85th percentiles based on all S and G-Experts' input for the Maine Yankee site.

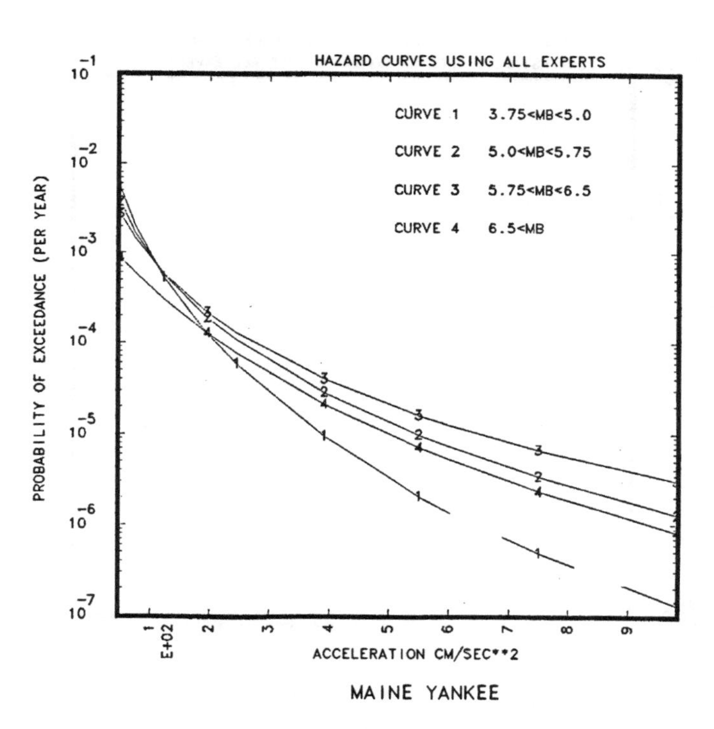

Figure 2.7.4 BEHCs which include only the contribution to the PGA hazard from earthquakes within the indicated magnitude range for the Maine Yankee site.

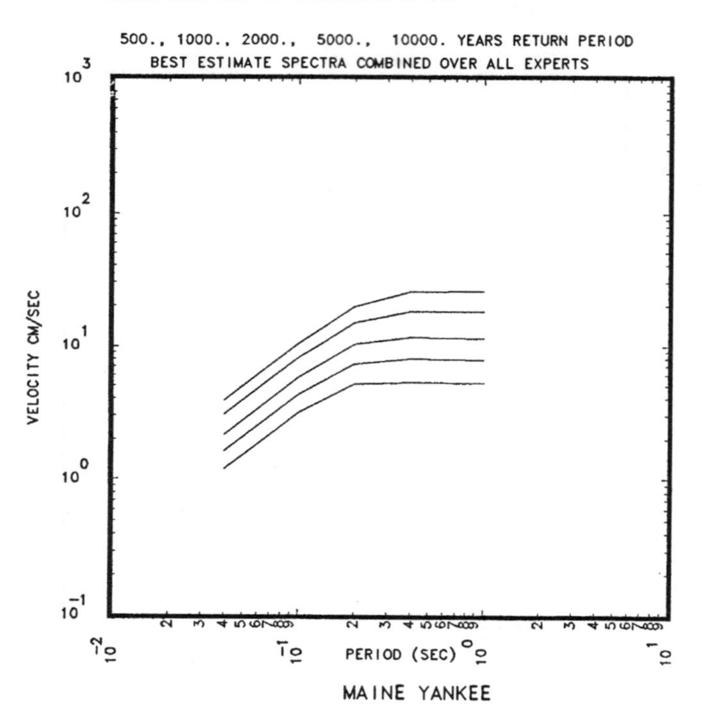

Figure 2.7.5 BEUHS for return periods of 500, 1000, 2000, 5000 and 10000 years aggregated over all S and G-Experts for the Maine Yankee site.

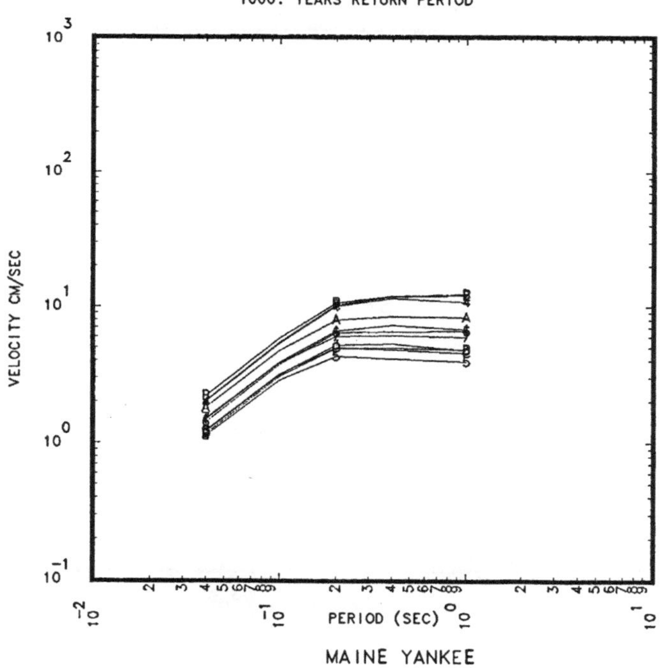

Figure 2.7.6 The 1000 year return period BEUHS per S-Expert aggregated over all G-Experts for the Maine Yankee site. Plot symbols are given in Table 2.0.

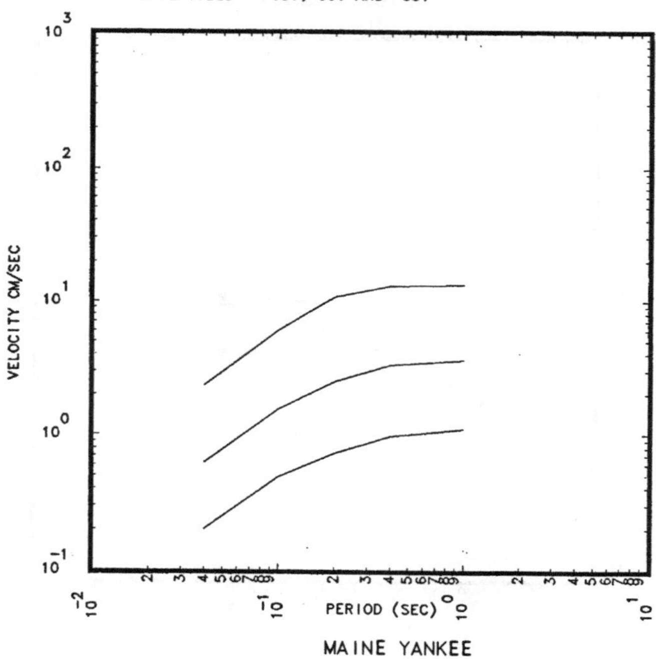

Figure 2.7.7 500 year return period CPUHS for the 15th, 50th and 85th percentiles aggregated over all S and G-Experts for the Maine Yankee site.

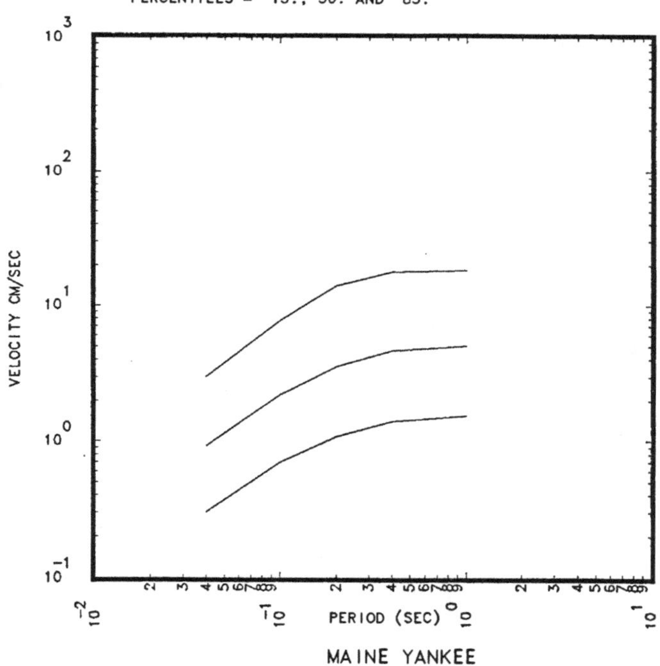

Figure 2.7.8 1000 year return period CPUHS for the 15th, 50th and 85th percentile aggregated over all S and G-Experts for the Maine Yankee site.

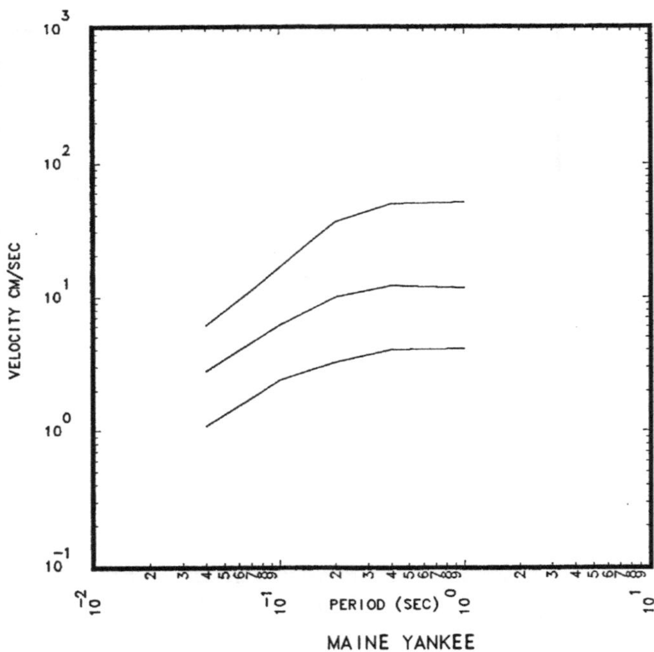

Figure 2.7.9 10000 year return period CPUHS for the 15th, 50th and 85th percentiles aggregated over all S and G-Experts for the Maine Yankee site.

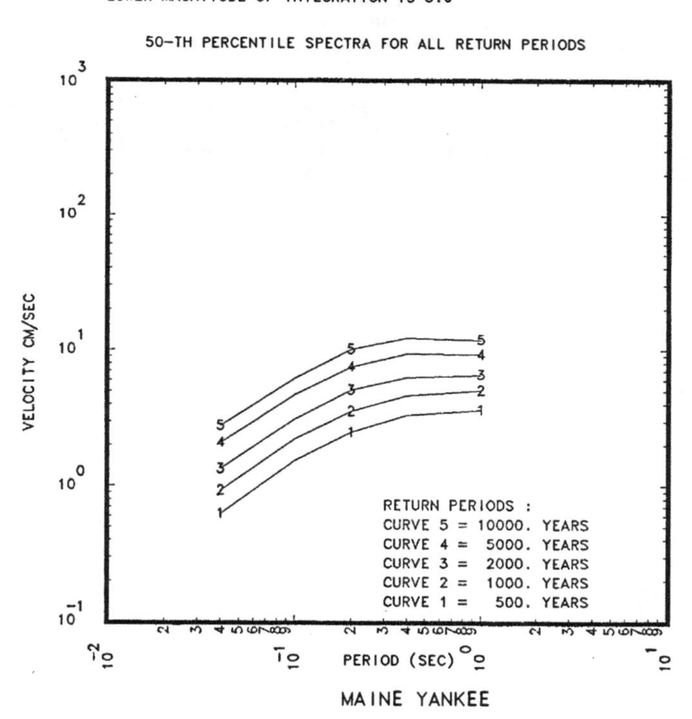

Figure 2.7.10 Comparison of the 50th percentile CPUHS for return periods of 500, 1000, 2000, 5000 and 10000 years for the Maine Yankee site.

2.8 MILLSTONE

Millstone is a rock site and is represented by the symbol "8" in Fig. 1.1. Table 2.8.1 and Figs. 2.8.1 to 2.8.10 give the basic results for the Millstone site. Millstone was one of the ten test sites analyzed in Bernreuter et al. (1985). In Vol. 1, Section 4-1 comparisons were made between the results plotted in this section and our previous results given in Bernreuter et al. (1985). We concluded in Vol. 1 that there was little change in the PGA hazard but there was significant differences in the CPUHS.

We see from Fig. 2.8.4 that if earthquakes in the range 3.75 to 5 were included, the hazard at the Millstone site would be increased by a factor 2 at 0.05g, a factor of 1.3 at 0.2g to very little at 0.6g.

We see from Table 2.8.1, that unlike at the Maine Yankee site previously discussed, the zone that contains the site is the most significant contributor to the BEHC for most S-Experts. Thus, the spread between the G-Experts BEHCs per S-Expert is similar to that shown in Fig. 2.1.11.

TABLE 2.8.1

MOST IMPORTANT ZONES PER S-EXPERT
FOR MILLSTONE

SITE SOIL CATEGORY ROCK

S-XPT NUM.	HOST ZONE		ZONES CONTRIBUTING MOST SIGNIFICANTLY TO THE PGA BEHC MD % OF CONTRIBUTION AT LOW PGA(0.125G)				AT HIGH PGA(0.60G)			
1	ZONE 22	ZONE ID:	ZONE 22	ZONE 20	ZONE 21	ZONE 4	ZONE 22	ZONE 1	ZONE 20	ZONE 21
		% CON T:	56.	22.	13.	4.	87.	6.	5.	3.
2	ZONE 31	ZONE ID:	ZONE 31	ZONE 32	ZONE 28	COMP. ZON	ZONE 31	ZONE 32	COMP. ZON	ZONE 28
		% CONT.:	67.	23.	6.	3.	86.	11.	2.	0.
3	ZONE 4	ZONE ID:	ZONE 4	ZONE 3	ZONE 2	ZONE 5	ZONE 4	ZONE 2	ZONE 3	ZONE 5
		% CON T:	89.	4.	3.	3.	9.	0.	0.	0.
4	ZONE 23	ZONE ID:	ZONE 23	ZONE 18	ZONE 16	ZONE 19	ZONE 23	ZONE 18	ZONE 16	ZONE 20
		% CONT.:	39.	26.	14.	8.	83.	10.	4.	1.
5	ZONE 1	ZONE ID:	ZONE 1	ZONE 6	ZONE 3	ZONE 8	ZONE 1	ZONE 3	ZONE 8	ZONE 4
		% CON T:	77.	14.	5.	3.	99.	1.	0.	0.
6	ZONE 6	ZONE ID:	ZONE 6	ZONE 7	ZONE 5	ZONE 3	ZONE 6	ZONE 5	ZONE 7	ZONE 3
		% CON T:	75.	8.	8.	8.	95.	3.	2.	1.
7	ZONE 15	ZONE ID:	ZONE 19	ZONE 24	ZONE 17	ZONE 14	ZONE 24	ZONE 19	ZONE 15	ZONE 29
		% CON T:	31.	17.	12.	8.	45.	27.	13.	6.
10	ZONE 2	ZONE ID:	ZONE 2	ZONE 4A	ZONE 22	ZONE 23	ZONE 2	ZONE 4A	ZONE 22	ZONE 19 =
		% CONT.:	44.	13.	10.	8.	61.	27.	4.	4.
11	ZONE 1	ZONE ID:	ZONE 1	ZONE 3	ZONE 5	CZ = ZONE	ZONE 1	ZONE 5	ZONE 3	CZ = ZONE
		% CONT.:	38.	31.	18.	7.	65.	20.	9.	4.
12	ZONE 32	ZONE ID:	ZONE 32	ZONE 31	ZONE 34	ZONE 33	ZONE 32	ZONE 34	ZONE 37	ZONE 33
		% CONT.:	73.	9.	9.	5.	9.	1.	0.	0.
13	ZONE 10	ZONE ID:	ZONE 10	ZONE 12	CZ 15	ZONE 11	ZONE 10	CZ 15	ZONE 12	CZ 18
		% CONT.:	86.	7.	3.	2.	98.	2.	0.	0.

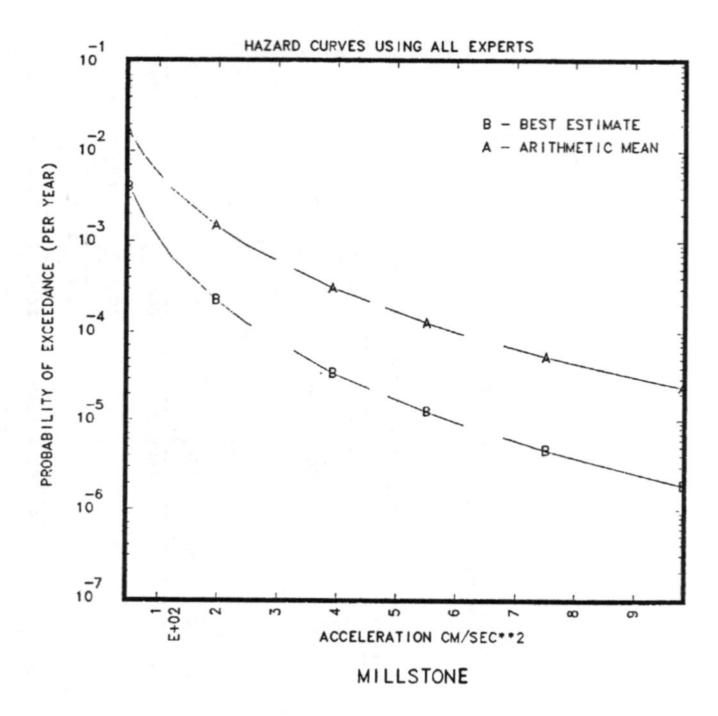

Figure 2.8.1 Comparison of the BEHC and AMHC aggregated over all S and G-Experts for the Millstone site.

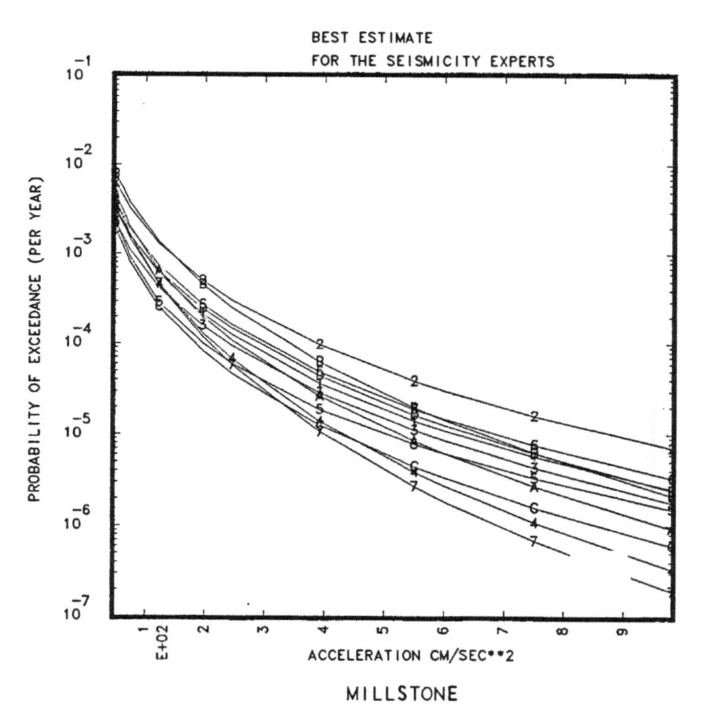

Figure 2.8.2 BEHCs per S-Expert combined over all G-Experts for the Mill Stone site. Plot symbols given in Table 2.0.

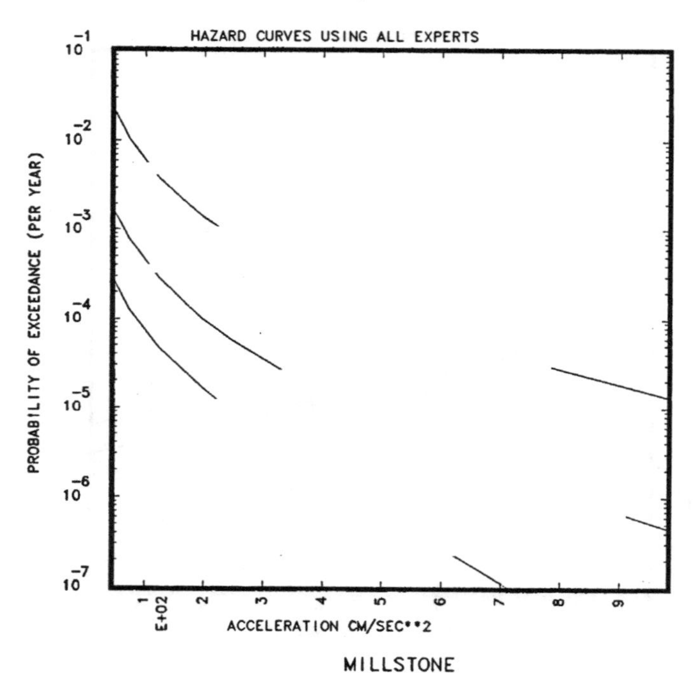

Figure 2.8.3 CPHCs for the 15th, 50th and 85th percentiles based on all S and G-Experts' input for the Mill Stone site.

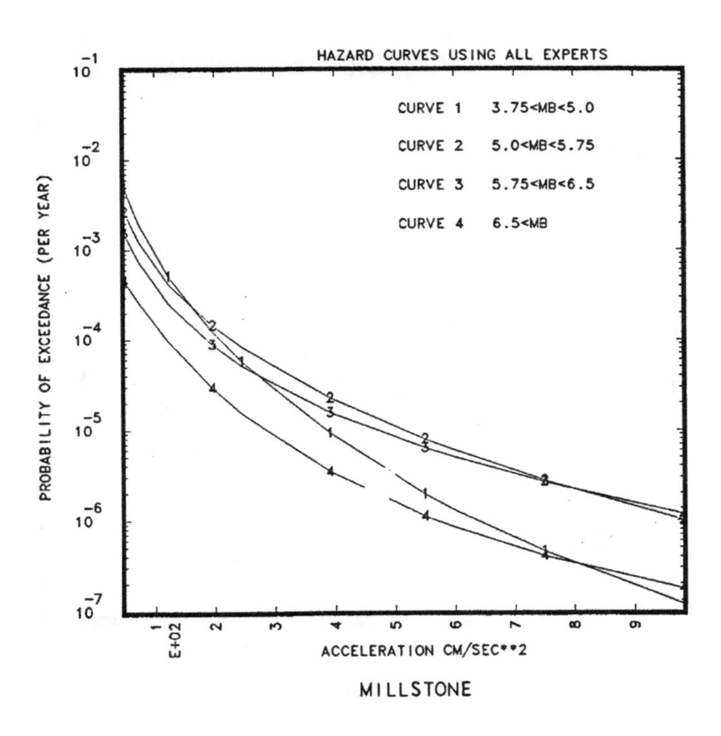

Figure 2.8.4 BEHCs which include only the contribution to the PGA hazard from earthquakes within the indicated magnitude range for the Mill Stone site.

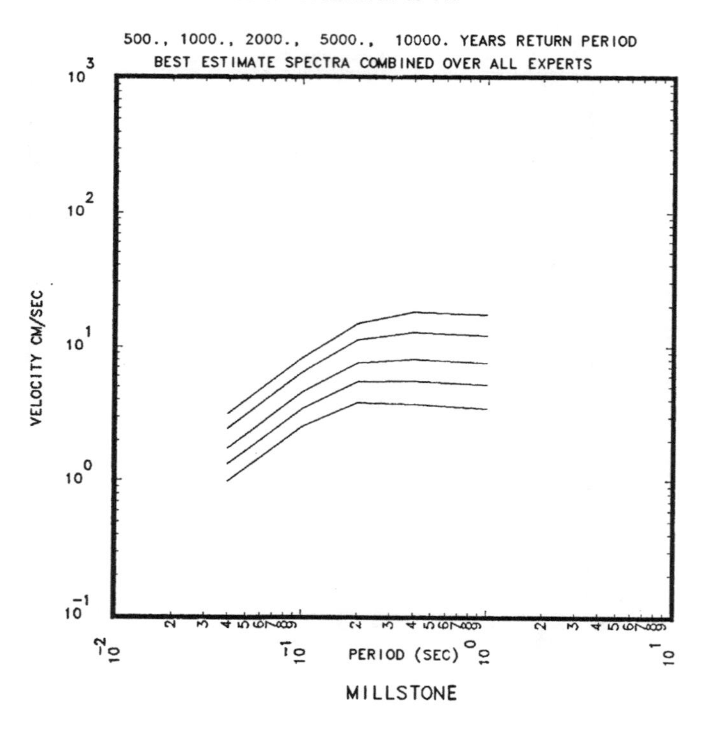

Figure 2.8.5 BEUHS for return periods of 500, 1000, 2000, 5000 and 10000 years aggregated over all S and G-Experts for the Mill Stone site.

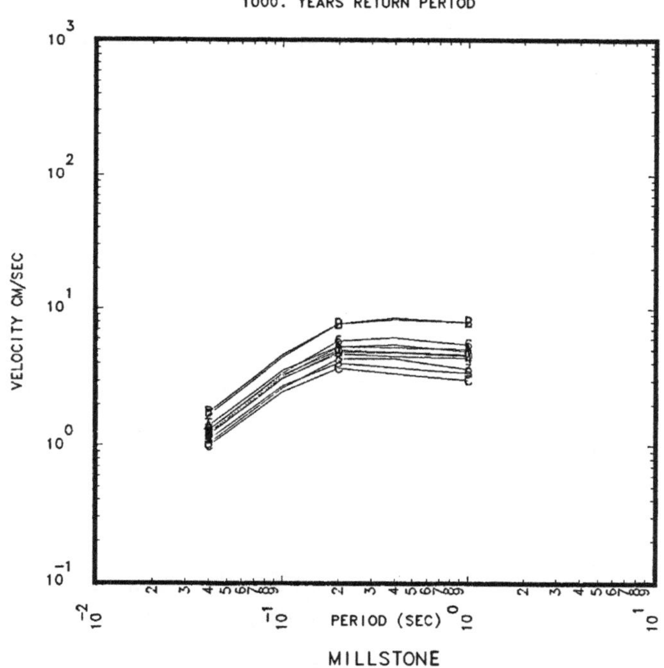

Figure 2.8.6 The 1000 year return period BEUHS per S-Expert aggregated over all G-Experts for the Mill Stone site. Plot symbols are given in Table 2.0.

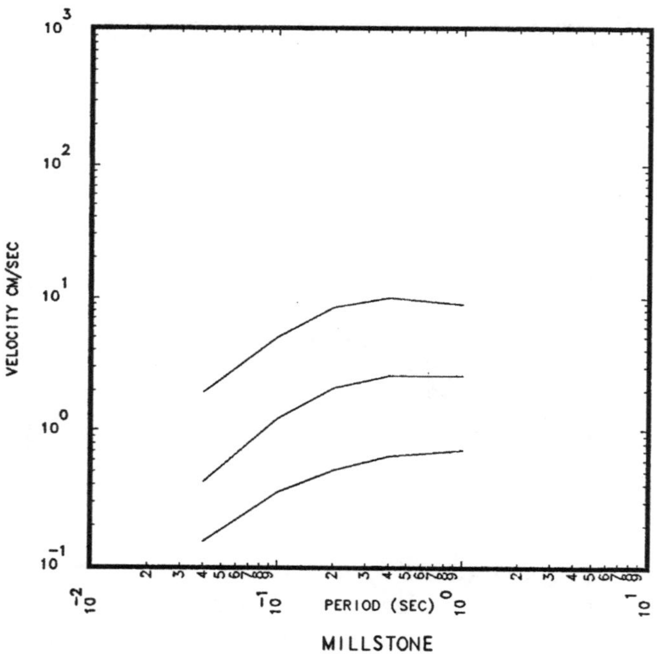

Figure 2.8.7 500 year return period CPUHS for the 15th, 50th and 85th percentiles aggregated over all S and G-Experts for the Mill Stone site.

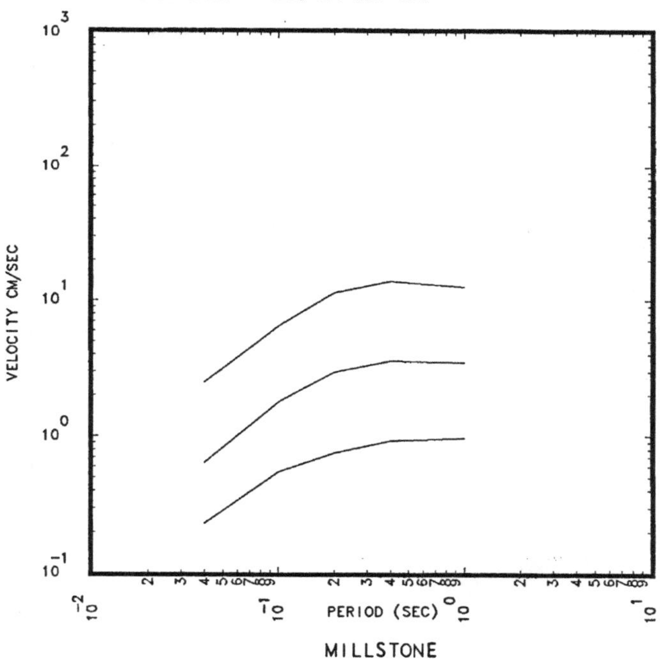

Figure 2.8.8 1000 year return period CPUHS for the 15th, 50th and 85th percentile aggregated over all S and G-Experts for the Mill Stone site.

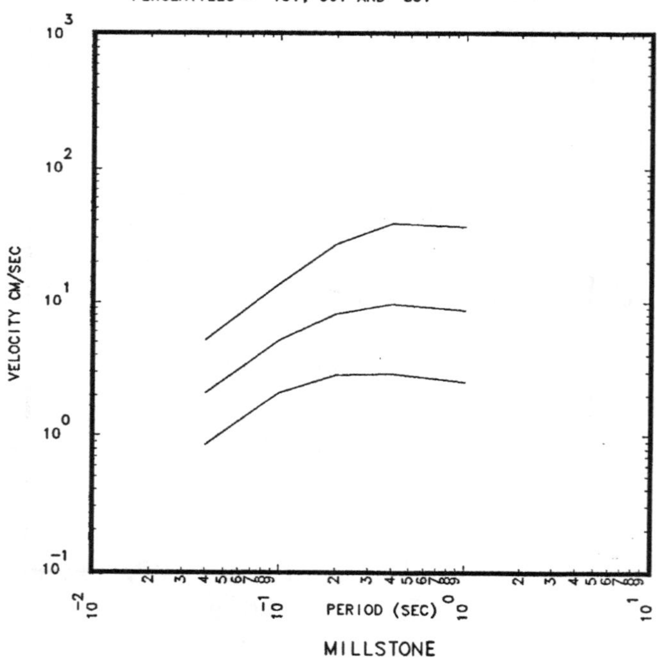

Figure 2.8.9 10000 year return period CPUHS for the 15th, 50th and 85th percentiles aggregated over all S and G-Experts for the Mill Stone site.

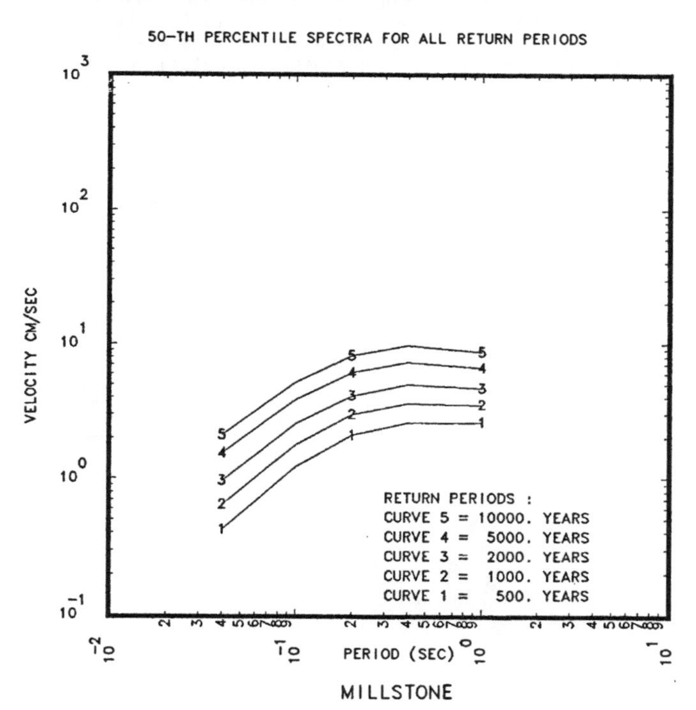

Figure 2.8.10 Comparison of the 50th percentile CPUHS for return Periods of 500, 1000, 2000, 5000 and 10000 years for the Mill Stone site.

2.9 NINE MILE POINT

Nine Mile Point is a rock site represented by the symbol "9" in Fig. 1.1. Table 2.9.1 and Figs. 2.9.1 to 2.9.10 give the basic results for the Nine Mile Point site.

It can be seen from Fig. 1.1 that the Nine Mile Point site is almost at the same location as the Fitzpatrick site. The results for the Nine Mile Point site are essentially the same as for the Fitzpatrick site, hence the discussion of Section 2.1 is not repeated. Because the uncertainty in the hazard is so large at the Nine Mile Point and Fitzpatrick sites , e.g., as measured by the large spread between the BEHC and AMHC shown in Fig. 2.9.1 or Fig. 2.1.1 there are differences between the results for the two sites due to the fact that the random sampling of the distributions are different for each Monte Carlo analysis. This point is discussed in detail in Section 3.

TABLE 2.9.1

MOST IMPORTANT ZONES PER S-EXPERT
FOR NINE MILE POINT

SITE SOIL CATEGORY ROCK

S-XPT NUM.	HOST ZONE		ZONES CONTRIBUTING MOST SIGNIFICANTLY TO THE PGA BEHC ND % OF CONTRIBUTION AT OW PGA(0.125G)				AT HIGH PGA(0.60G)			
1	ZONE 15	ZONE ID:	ZONE 20	ZONE 21	ZONE 19	ZONE 4	ZONE 20	ZONE 19	ZONE 21	ZONE 1
		% CON T.:	71.	11.	9.	4.	85.	10.	4.	2.
2	COMP. ZO	ZONE ID:	ZONE 32	ZONE 31	COMP. ZON	ZONE 28	ZONE 32	COMP. ZON	ZONE 31	ZONE 2
		% CONT.:	66.	17.	14.	4.	74.	23.	3.	0.
3	ZONE 11	ZONE ID:	ZONE 11	ZONE 2	ZONE 3	ZONE 4	ZONE 11	ZONE 2	ZONE 3	ZONE 5
		% CON T:	63.	27.	5.	2.	84.	16.	0.	0.
4	ZONE 13	ZONE ID:	ZONE 16	ZONE 19	ZONE 15	ZONE 18	ZONE 16	ZONE 13	ZONE 19	ZONE 1
		% CONT.:	78.	7.	4.	4.	98.	1.	0.	0.
5	COMP. ZO	ZONE ID:	ZONE 6	ZONE 4	ZONE 5	ZONE 3	ZONE 6	ZONE 5	ZONE 4	ZONE
		% CON T:	96.	1.	1.	1.	95.	2.	2.	1.
6	COMP. ZO	ZONE ID:	ZONE 7	ZONE 3	ZONE 8	ZONE 6	ZONE 7	COMP. ZON	ZONE 3	ZONE 8
		% CON T:	75.	14.	5.	3.	94.	2.	2.	1.
7	ZONE 41	ZONE ID:	ZONE 17	ZONE 18	ZONE 41	ZONE 26	ZONE 17	ZONE 41	ZONE 18	ZONE 2
		% CONT.:	56.	24.	11.	5.	70.	19.	10.	1.
10	ZONE 19	ZONE ID:	ZONE 19 =	ZONE 6	ZONE 7	ZONE 9	ZONE 19 =	ZONE 6	ZONE 7	ZONE
		% CONT.:	32.	29.	20.	8.	68.	25.	5.	2.
11	CZ = ZON	ZONE ID:	ZONE 3	ZONE 4	CZ = ZONE	ZONE 2	ZONE 4	CZ = ZONE	ZONE 3	ZONE
		% CON T:	43.	19.	18.	8.	47.	27.	23.	2.
12	ZONE 4 =	ZONE ID:	ZONE 31	ZONE 30	ZONE 34	ZONE 31A	ZONE 31	ZONE 30	ZONE 34	ZONE 3
		% CON T.:	72.	15.	9.	2.	76.	18.	6.	0.
13	CZ 15	ZONE ID:	ZONE 11	CZ 15	ZONE 12	ZONE 10	CZ 15	ZONE 11	ZONE 12	ZONE
		% CONT.:	41.	34.	22.	3.	83.	12.	5.	0.

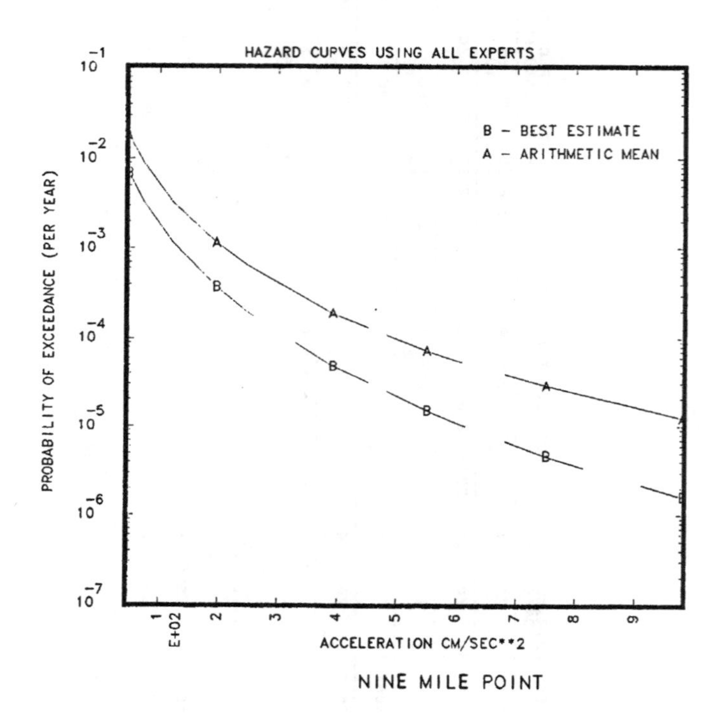

Figure 2.9.1 Comparison of the BEHC and AMHC aggregated over all S and G-Experts for the Nine Mile Point site.

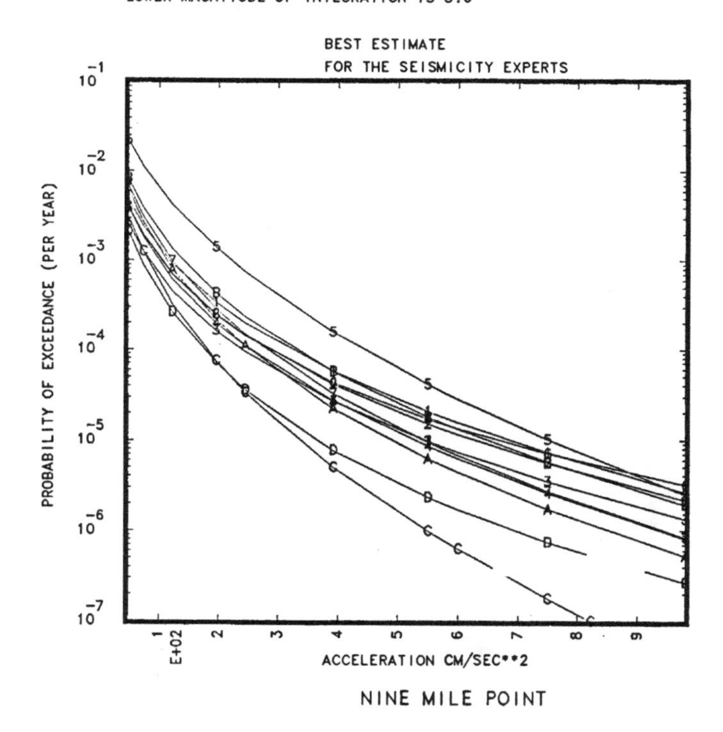

Figure 2.9.2 BEHCs per S-Expert combined over all G-Experts for the Nine Mile Pt. site. Plot symbols given in Table 2.0.

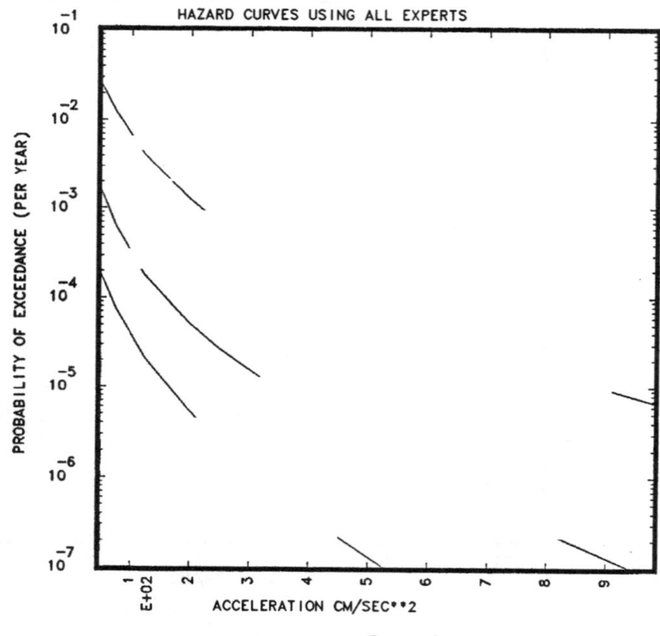

Figure 2.9.3 CPHCs for the 15th, 50th and 85th percentiles based on all S and G-Experts' input for the Nine Mile Pt. site.

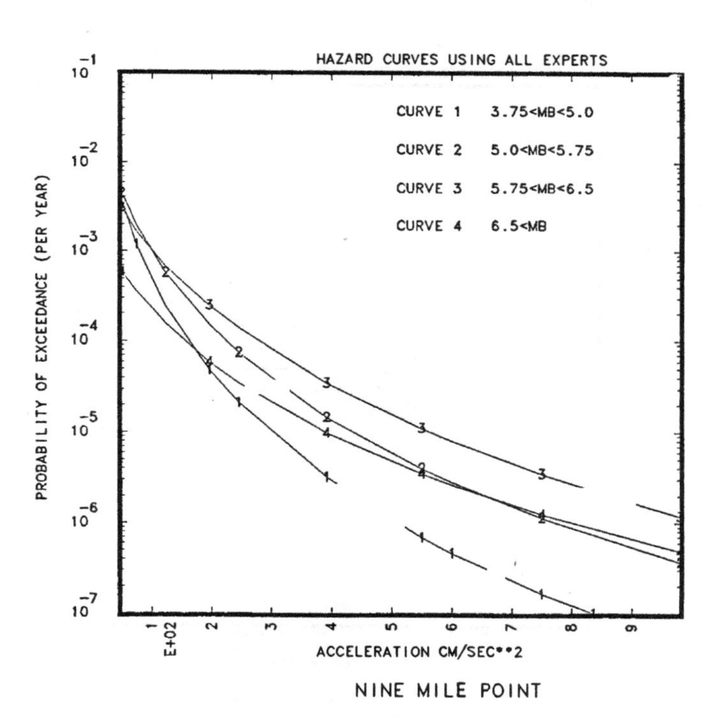

Figure 2.9.4 BEHCs which include only the contribution to the PGA hazard from earthquakes within the indicated magnitude range for the Nine Mile Pt. site.

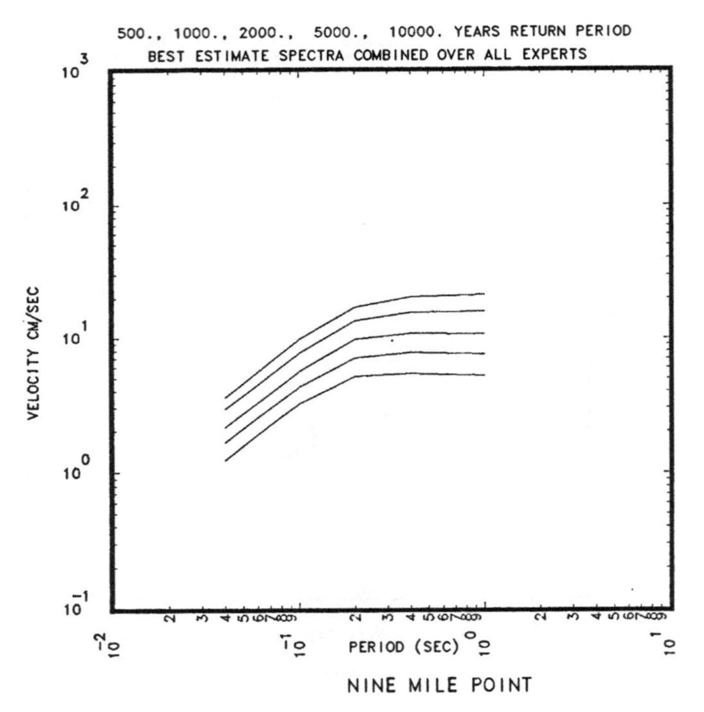

Figure 2.9.5 BEUHS for return periods of 500, 1000, 2000, 5000 and 10000 years aggregated over all S and G-Experts for the Nine Mile Pt. site.

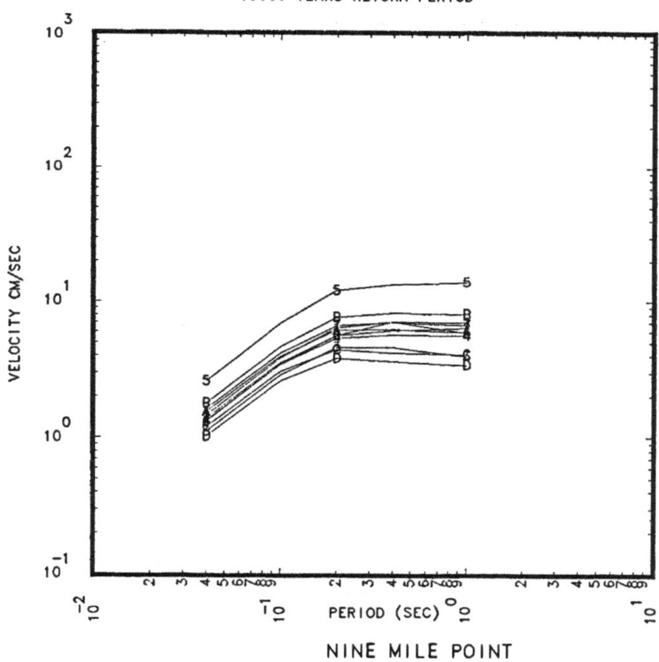

Figure 2.9.6 The 1000 year return period BEUHS per S-Expert aggregated over all G-Experts for the Nine Mile Pt. site. Plot symbols are given in Table 2.0.

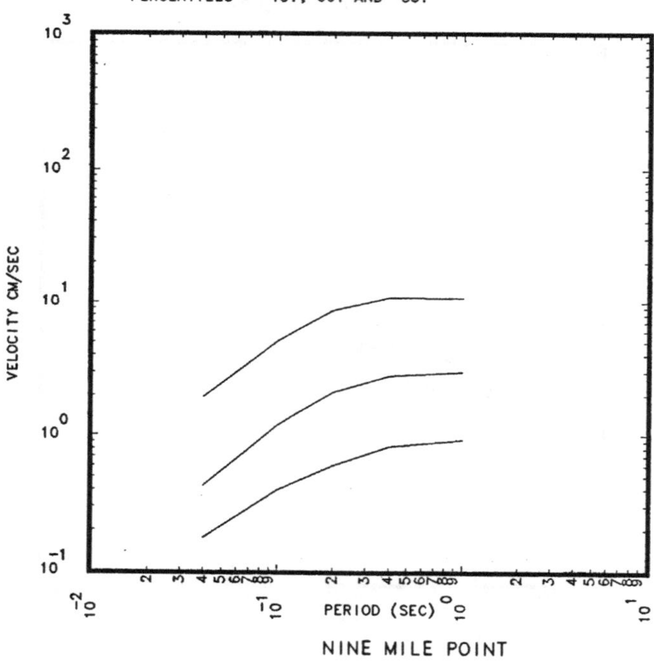

Figure 2.9.7 500 year return period CPUHS for the 15th, 50th and 85th percentiles aggregated over all S and G-Experts for the Nine Mile Pt. site.

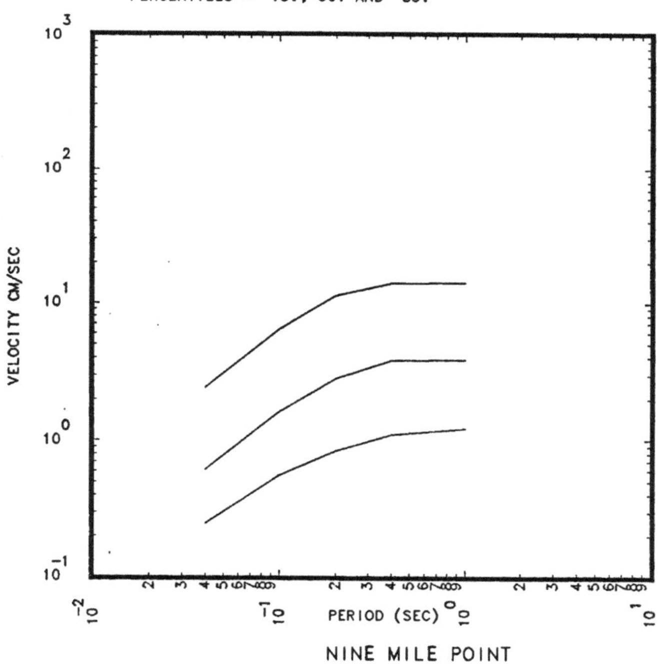

Figure 2.9.8 1000 year return period CPUHS for the 15th, 50th and 85th percentile aggregated over all S and G-Experts for the Nine Mile Pt. site.

E.U.S SEISMIC HAZARD CHARACTERIZATION
LOWER MAGNITUDE OF INTEGRATION IS 5.0

10000.-YEAR RETURN PERIOD CONSTANT PERCENTILE SPECTRA FOR :

PERCENTILES = 15., 50. AND 85.

VELOCITY CM/SEC

PERIOD (SEC)

NINE MILE POINT

Figure 2.9.9 10000 year return period CPUHS for the 15th, 50th and 85th percentiles aggregated over all S and G-Experts for the Nine Mile Pt. site.

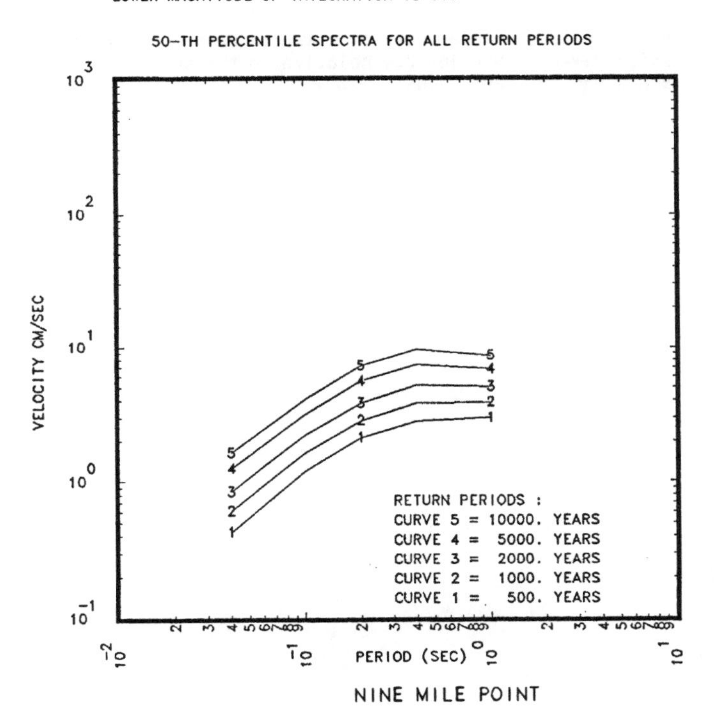

Figure 2.9.10 Comparison of the 50th percentile CPUHS for return periods of 500, 1000, 2000, 5000 and 10000 years for the Nine Mile Pt. site.

2.10 OYSTER CREEK

Oyster Creek is a deep soil site represented by the symbol "A" in Fig. 1.1. Table 2.10.1 and Figs. 2.10.1 to 2.10.10 give the basic results for the Oyster Creek site. The AMHC is about the same as the 85th percentile CPHC.

We see from Fig. 2.10.4 that if earthquakes in the magnitude range 3.75 to 5.0 are included the PGA hazard would be increased by about a factor of 2 at 0.05g, a factor of 1.3 at 0.2g, and a factor of about 1.0 by about 0.5g.

Oyster Creek is near the Hope Creek site as can be seen from Fig. 1.1. Thus the discussion given in Section 2.4 relative to the spread between the G-Experts' BEHCs per S-Expert holds.

TABLE 2.10.1

MOST IMPORTANT ZONES PER S-EXPERT FOR OYSTER CREEK

SITE SOIL CATEGORY DEEP-SOIL

S-XPT NUM.	HOST ZONE		ZONES CONTRIBUTING MOST SIGNIFICANTLY TO THE BA BEHC ND % OF CONTRIBUTION AT LOW PGA(0.125G)				AT HIGH PGA(0.60G)			
1	ZONE 1	ZONE ID:	ZONE 1	ZONE 4	ZONE 20	ZONE 22	ZONE 1	ZONE 4	ZONE 22	ZONE 2
		% CONT.:	65.	18.	9.	4.	98.	2.	0.	0.
2	ZONE 28	ZONE ID:	ZONE 28	ZONE 32	ZONE 31	COMP. ZON	ZONE 28	COMP. ZON	ZONE 32	ZONE 3
		% CONT.:	94.	2.	2.	1.	99.	1.	0.	0.
3	ZONE 8A	ZONE ID:	ZONE 8A	ZONE 5	ZONE 4	COMP. ZON	ZONE 8A	ZONE 5	ZONE 4	COMP.
		% CONT.:	74.	17.	7.	1.	97.	2.	1.	0.
4	COMP. ZO	ZONE ID:	ZONE 11	ZONE 16	COMP. ZON	ZONE 12	COMP. ZON	ZONE 11	ZONE 7	ZONE 2
		% CONT.:	31.	22.	19.	8.	88.	12.	0.	0.
5	ZONE 8	ZONE ID:	ZONE 8	ZONE 1	ZONE 9	ZONE 6	ZONE 8	ZONE 1	COMP. ZON	ZONE
		% CONT.:	51.	41.	4.	3.	54.	45.	0.	0.
6	ZONE 6	ZONE ID:	ZONE 6	ZONE 7	ZONE 3	ZONE 13	ZONE 6	ZONE 7	COMP. ZON	ZONE 3
		% CONT.:	98.	1.	0.	0.	100.	0.	0.	0.
7	ZONE 29	ZONE ID:	ZONE 29	ZONE 14	ZONE 13	ZONE 7	ZONE 29	ZONE 14	ZONE 24	ZONE 1
		% CONT.:	63.	28.	6.	2.	87.	11.	1.	1.
10	ZONE 4B	ZONE ID:	ZONE 4B	ZONE 19 =	ZONE 5	ZONE 4A	ZONE 4B	ZONE 19 =	ZONE 1	ZONE
		% CONT.:	94.	3.	2.	0.	99.	1.	0.	0.
11	CZ = ZON	ZONE ID:	ZONE 5	CZ = ZONE	ZONE 3	ZONE 4	ZONE 5	CZ = ZONE	ZONE 8	ZONE
		% CONT.:	53.	38.	6.	2.	51.	49.	0.	0.
12	ZONE 32	ZONE ID:	ZONE 32	ZONE 25	ZONE 31	ZONE 34	ZONE 32	ZONE 25	ZONE 19	ZONE 2
		% CONT.:	98.	1.	0.	0.	100.	0.	0.	0.
13	CZ 17	ZONE ID:	CZ 17	ZONE 10	CZ 15	ZONE 12	CZ 17	ZONE 10	CZ 15	ZONE
		% CONT.:	84.	12.	2.	1.	98.	1.	1.	0.

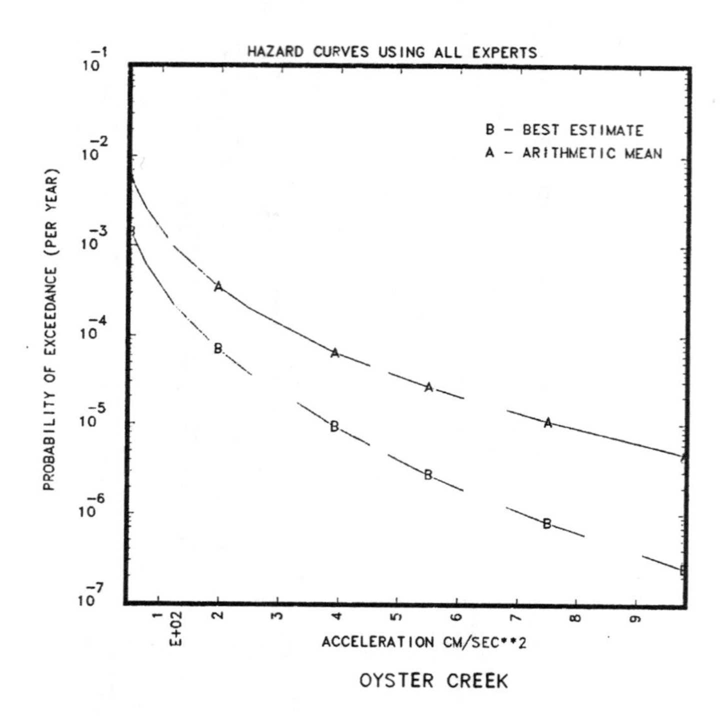

Figure 2.10.1 Comparison of the BEHC and AMHC aggregated over all S and G-Experts for the Oyster Creek site.

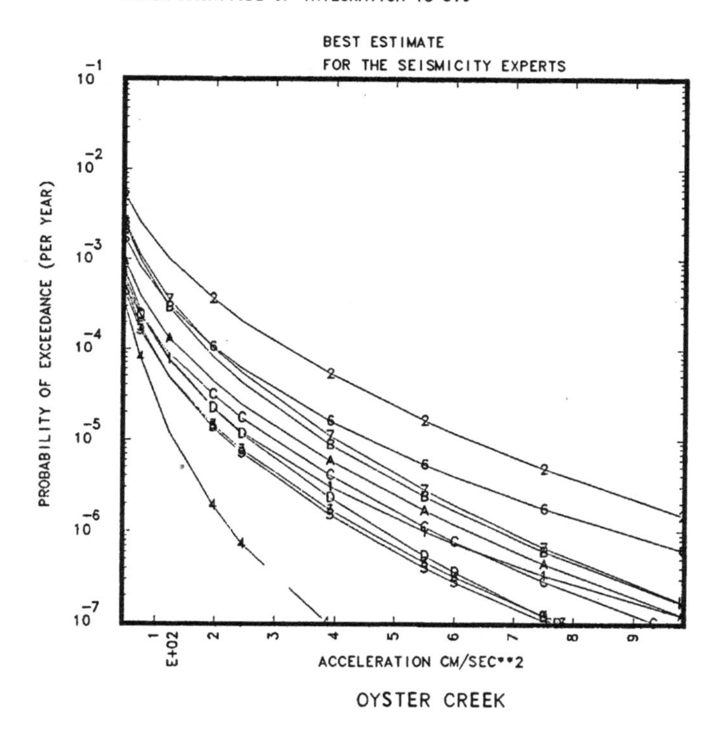

Figure 2.10.2 BEHCs per S-Expert combined over all G-Experts for the Oyster Creek site. Plot symbols given in Table 2.0.

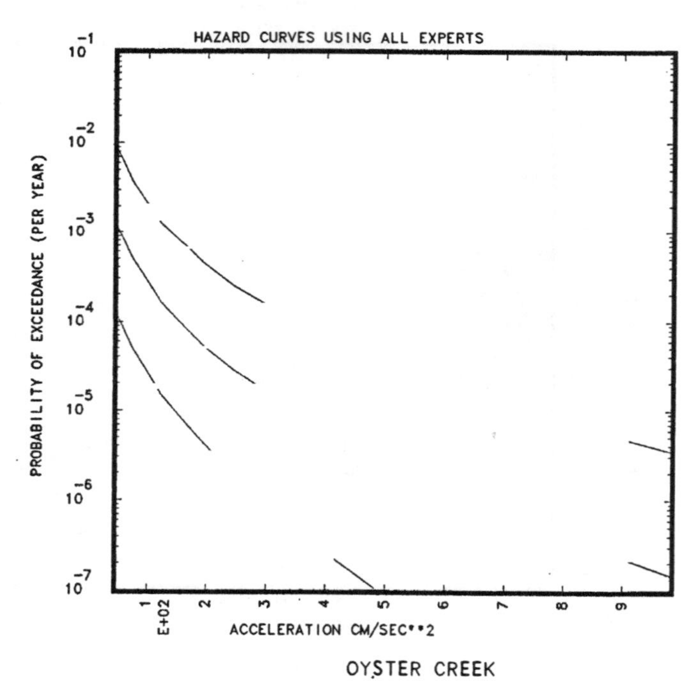

Figure 2.10.3 CPHCs for the 15th, 50th and 85th percentiles based on all S and G-Experts' input for the Oyster Creek site.

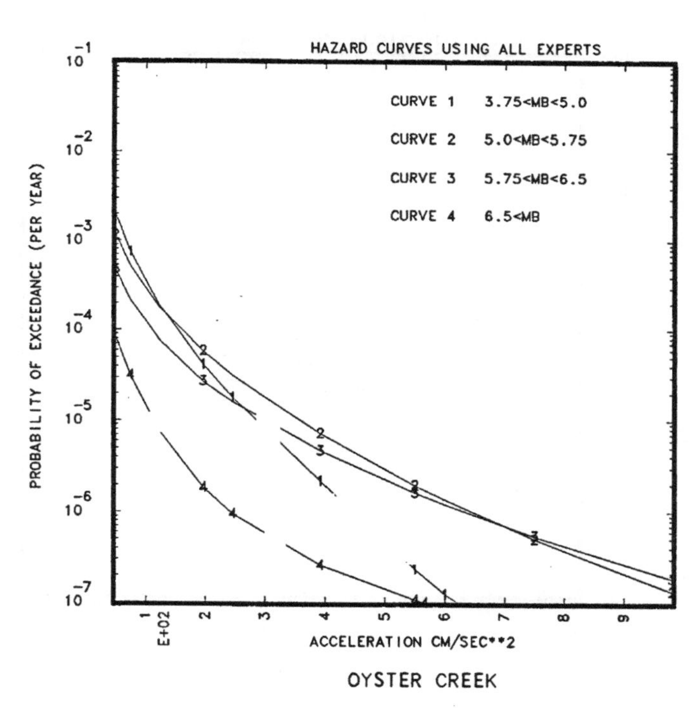

Figure 2.10.4 BEHCs which include only the contribution to the PGA hazard from earthquakes within the indicated magnitude range for the Oyster Creek site.

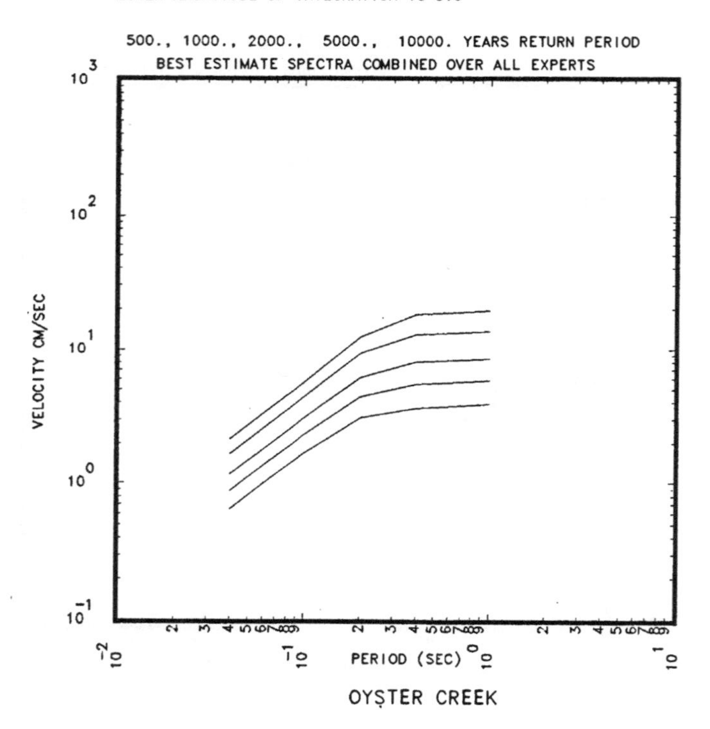

Figure 2.10.5 BEUHS for return periods of 500, 1000, 2000, 5000 and 10000 years aggregated over all S and G-Experts for the Oyster Creek site.

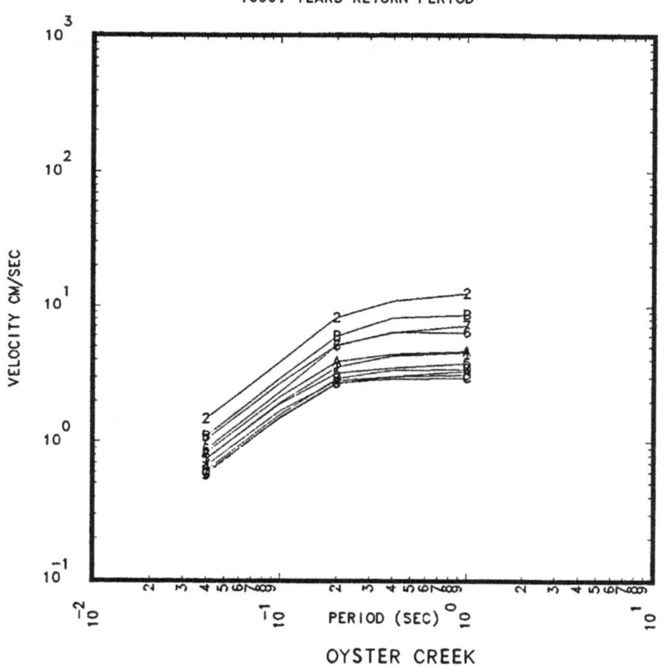

Figure 2.10.6 The 1000 year return period BEUHS per S-Expert aggregated over all G-Experts for the Oyster Creek site. Plot symbols are given in Table 2.0.

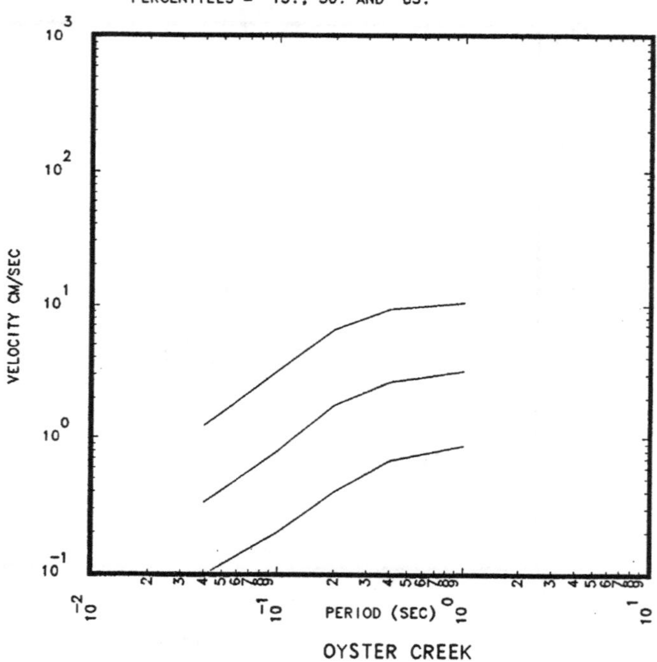

Figure 2.10.7 500 year return period CPUHS for the 15th, 50th and 85th percentiles aggregated over all S and G-Experts for the Oyster Creek site.

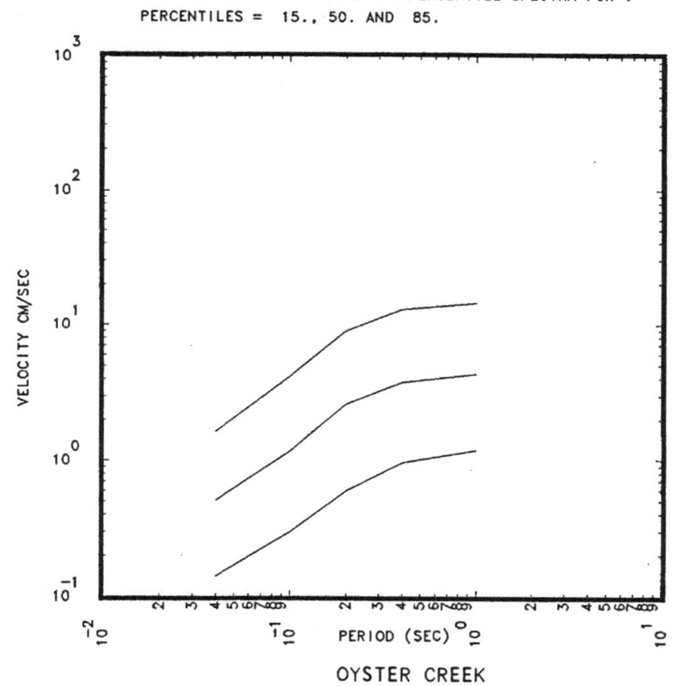

Figure 2.10.8 1000 year return period CPUHS for the 15th, 50th and 85th percentile aggregated over all S and G-Experts for the Oyster Creek site.

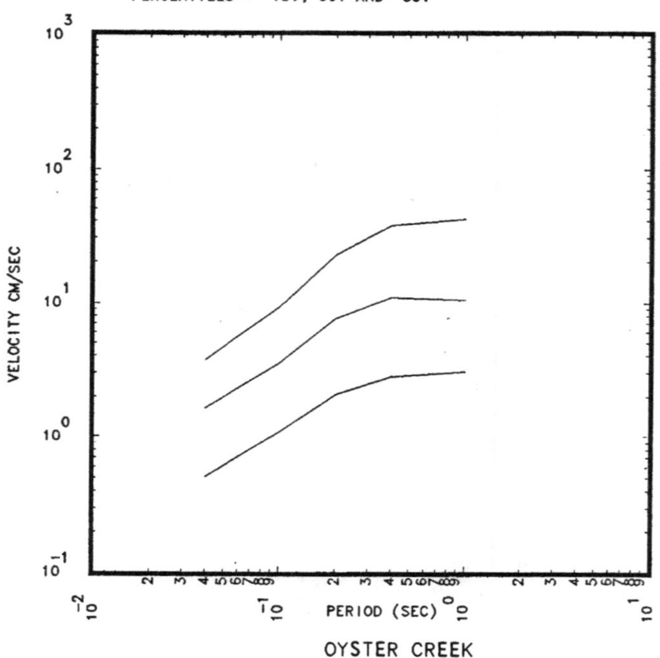

Figure 2.10.9 10000 year return period CPUHS for the 15th, 50th and 85th percentiles aggregated over all S and G-Experts for the Oyster Creek site.

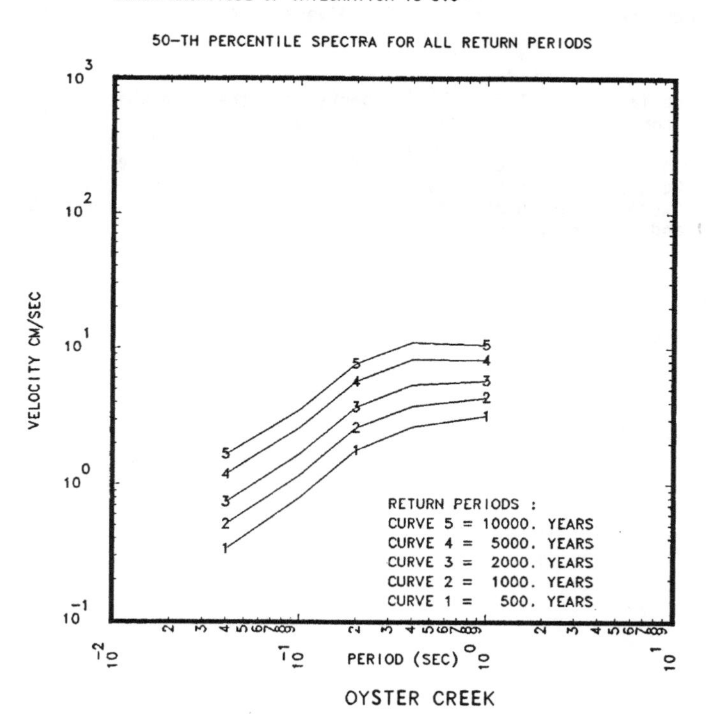

Figure 2.10.10 Comparison of the 50th percentile CPUHS for return periods of 500, 1000, 2000, 5000 and 10000 years for the Oyster Creek site.

2.11 PEACH BOTTOM

The Peach Bottom site is a rock site represented by the symbol "B" in Fig. 1.1. Table 2.11.1 and Figs. 2.11.1 to 2.11.10 give the basic results for the Peach Bottom site. The hazard curves for the Peach Bottom site were developed using Region 1 GM models. As noted in Section 2.0 the Peach Bottom site could be considered as part of either Region 2 or 3. The sensitivity of our placing the Peach Bottom site in region is discussed in Section 3.

The AMHC is about the same as the 85th percentile CPHC. We see from Fig. 2.11.2 that there is a relatively wide diversity of opinion among the S-Experts.

We see from Table 2.11.1 for all S-Experts that the zone which contains the site also contribute most to the BEHC. Also, we see from Fig. 2.11.4 that earthquakes in the 3.75 to 5.0 magnitude range included in the PGA hazard curve would be increased by a factor of over 2.0 at 0.05g, a factor of about 1.3 at 0.2g and dropping to a factor of about 1.0 at about 0.5g. Also, the spread between the G-Experts' BEHCs per S-Expert is typical for rock sites in Region 1 and is similar to that shown in Fig. 2.1.11.

TABLE 2.11.1

MOST IMPORTANT ZONES PER S-EXPERT FOR PEACH BOTTOM

SITE SOIL CATEGORY ROCK

S-XPT NUM.	HOST ZONE		ZONES CONTRIBUTING MOST SIGNIFICANTLY TO THE PGA BEHC AND % OF CONTRIBUTION AT LOW PGA(0.125G)				AT HIGH PGA(0.60G)			
1	ZONE 4	ZONE ID:	ZONE 4	ZONE 1	ZONE 20	ZONE 3	ZONE 4	ZONE 1	ZONE 20	ZONE 3
		% CONT.:	80.	7.	6.	2.	93.	6.	0.	0.
2	ZONE 28	ZONE ID:	ZONE 28	ZONE 30	ZONE 32	ZONE 27	ZONE 28	ZONE 32	ZONE 30	ZONE 2
		% CONT.:	82.	7.	4.	4.	97.	1.	1.	1.
3	ZONE 5	ZONE ID:	ZONE 5	ZONE 8A	ZONE 4	ZONE 1	ZONE 5	ZONE 8A	COMP. ZON	ZONE 4
		% CONT.:	89.	5.	2.	1.	98.	1.	0.	0.
4	ZONE 11	ZONE ID:	ZONE 11	ZONE 12	ZONE 16	ZONE 8	ZONE 1	ZONE 12	ZONE 8	ZONE 1
		% CONT.:	43.	21.	12.	11.	88.	6.	6.	0.
5	ZONE 1	ZONE ID:	ZONE 1	ZONE 6	ZONE 9	ZONE 8	ZONE 1	ZONE 9	ZONE 8	ZONE
		% CONT.:	44.	39.	12.	2.	98.	1.	0.	0.
6	ZONE 6	ZONE ID:	ZONE 6	ZONE 7	ZONE 13	ZONE 15	ZONE 6	ZONE 7	ZONE 13	ZONE 1
		% CONT.:	85.	4.	4.	3.	99.	0.	0.	0.
7	ZONE 7	ZONE ID:	ZONE 7	ZONE 29	ZONE 13	ZONE 10	ZONE 7	ZONE 29	ZONE 13	ZONE 1
		% CONT.:	70.	17.	4.	3.	94.	6.	0.	0.
10	ZONE 4B	ZONE ID:	ZONE 4B	ZONE 5	ZONE 19 =	ZONE 6	ZONE 4B	ZONE 5	ZONE 19 =	ZONE
		% CONT.:	63.	26.	3.	2.	97.	2.	1.	0.
11	ZONE 5	ZONE ID:	ZONE 5	CZ = ZONE	ZONE 3	ZONE 8	ZONE 5	CZ = ZONE	ZONE 8	ZONE
		% CONT.:	81.	8.	5.	3.	97.	2.	0.	0.
12	ZONE 32	ZONE ID:	ZONE 32	ZONE 31	ZONE 27	ZONE 23A	ZONE 32	ZONE 27	ZONE 23A	ZONE 1
		% CONT.:	90.	4.	3.	1.	100.	0.	0.	0.
13	CZ 15	ZONE ID:	CZ 15	CZ 17	ZONE 10	ZONE 11	CZ 15	CZ 17	ZONE 10	ZONE
		% CONT.:	55.	28.	10.	2.	90.	9.	0.	0.

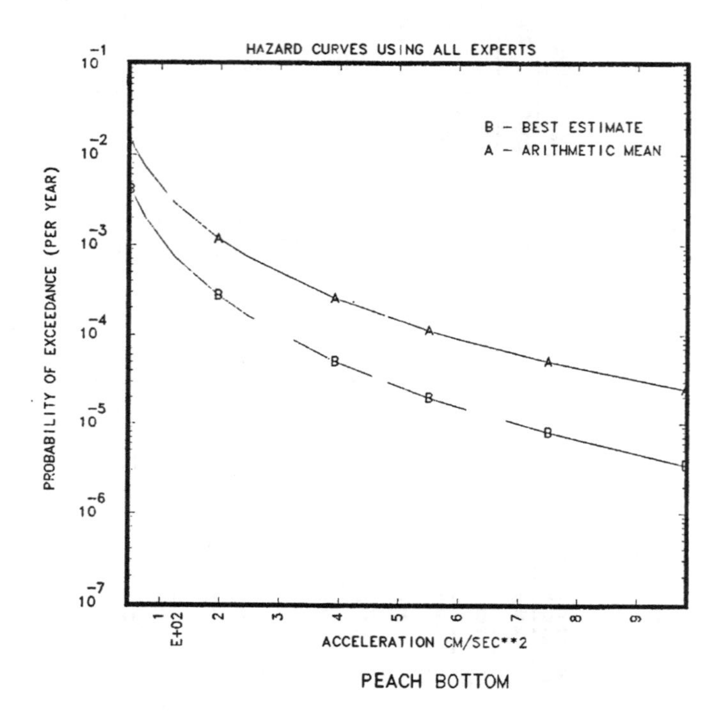

Figure 2.11.1 Comparison of the BEHC and AMHC aggregated over all S and G-Experts for the Peach Bottom site.

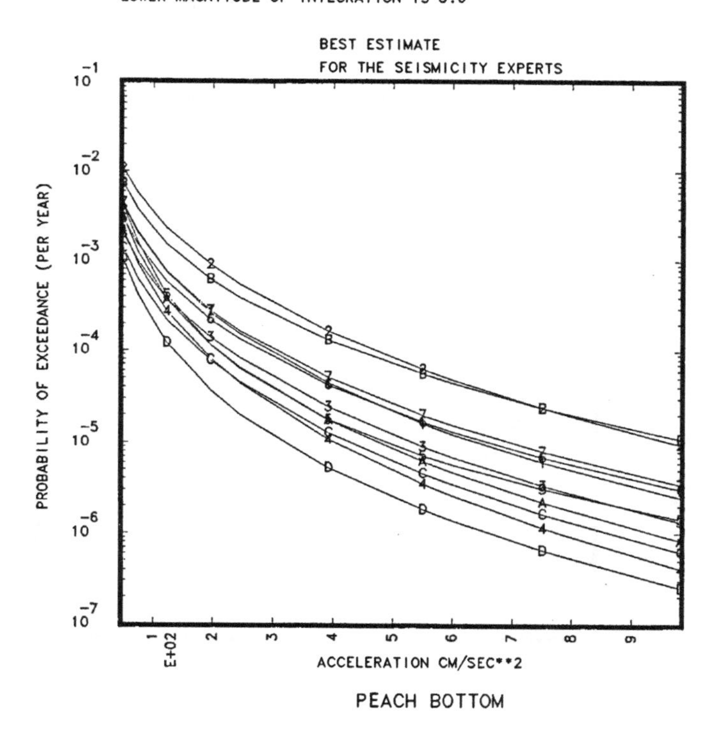

Figure 2.11.2 BEHCs per S-Expert combined over all G-Experts for the Peach Bottom site. Plot symbols given in Table 2.0.

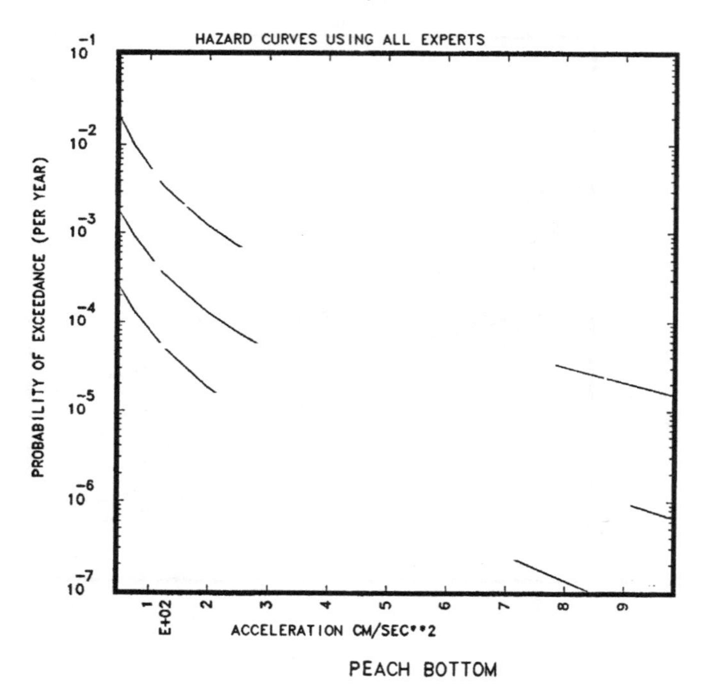

Figure 2.11.3 CPHCs for the 15th, 50th and 85th percentiles based on all S and G-Experts' input for the Peach Bottom site.

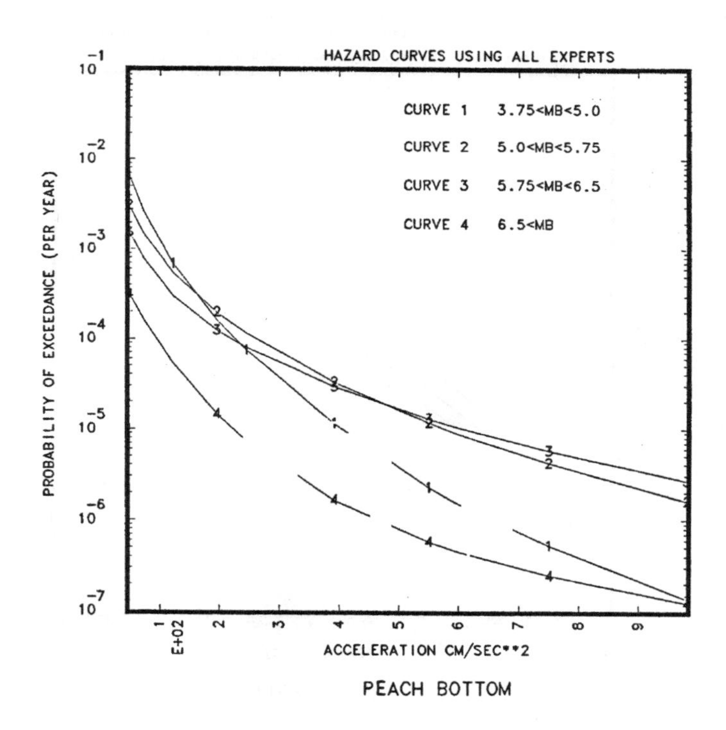

Figure 2.11.4 BEHCs which include only the contribution to the PGA hazard from earthquakes within the indicated magnitude range for the Peach Bottom site.

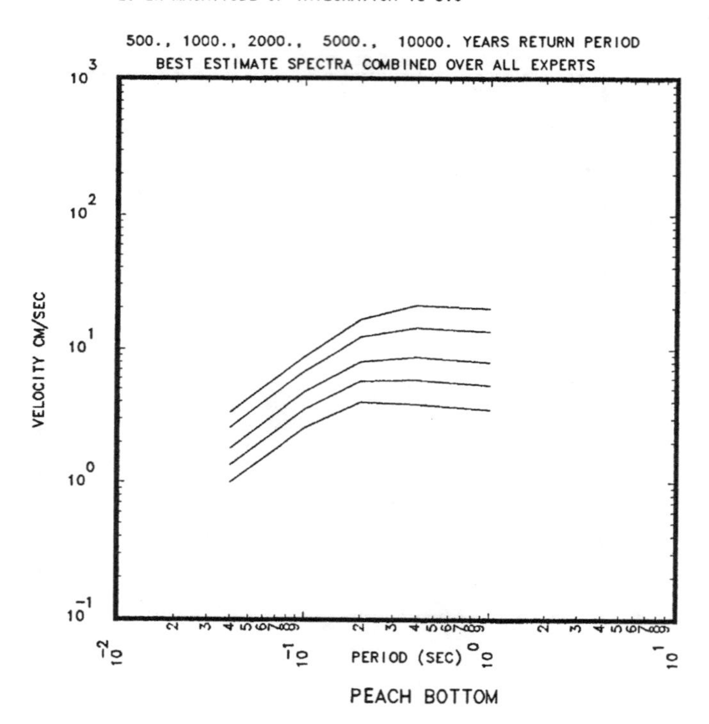

Figure 2.11.5 BEUHS for return periods of 500, 1000, 2000, 5000 and 10000 years aggregated over all S and G-Experts for the Peach Bottom site.

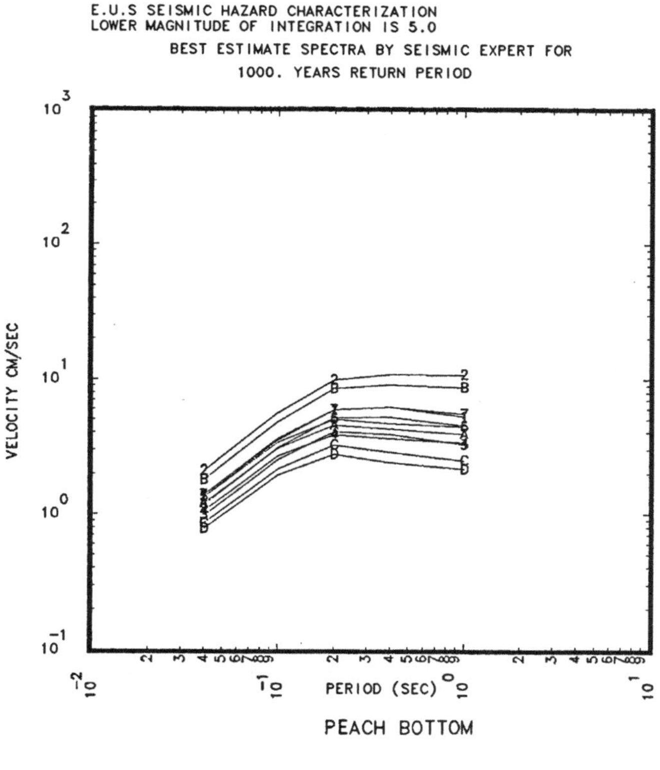

E.U.S SEISMIC HAZARD CHARACTERIZATION
LOWER MAGNITUDE OF INTEGRATION IS 5.0

BEST ESTIMATE SPECTRA BY SEISMIC EXPERT FOR
1000. YEARS RETURN PERIOD

VELOCITY CM/SEC

PERIOD (SEC)

PEACH BOTTOM

Figure 2.11.6 The 1000 year return period BEUHS per S-Expert aggregated over all G-Experts for the Peach Bottom site. Plot symbols are given in Table 2.0.

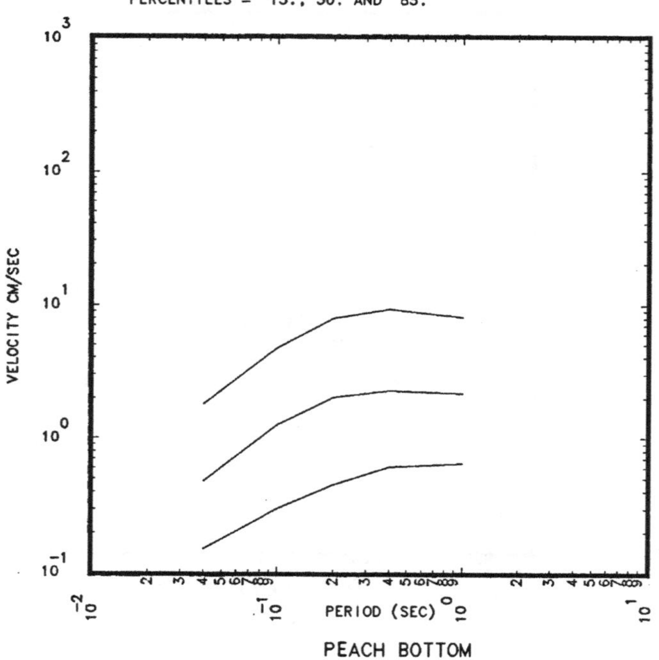

Figure 2.11.7 500 year return period CPUHS for the 15th, 50th and 85th percentiles aggregated over all S and G-Experts for the Peach Bottom site.

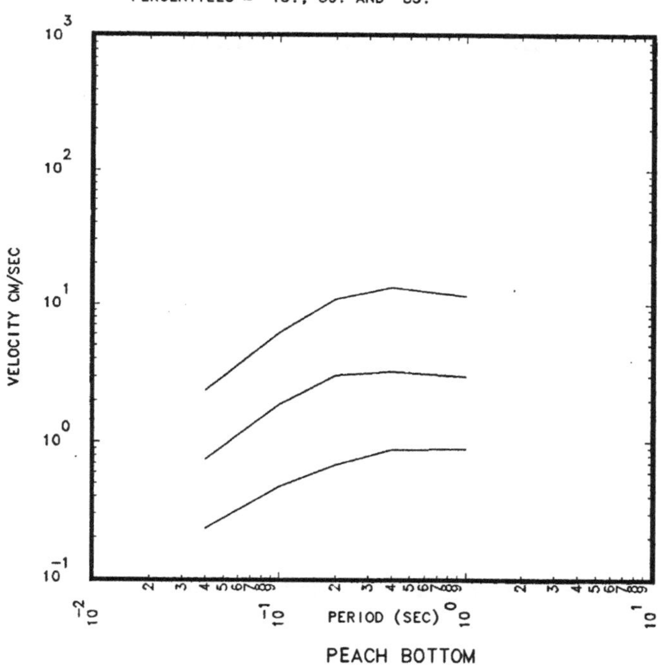

Figure 2.11.8 1000 year return period CPUHS for the 15th, 50th and 85th percentile aggregated over all S and G-Experts for the Peach Bottom site.

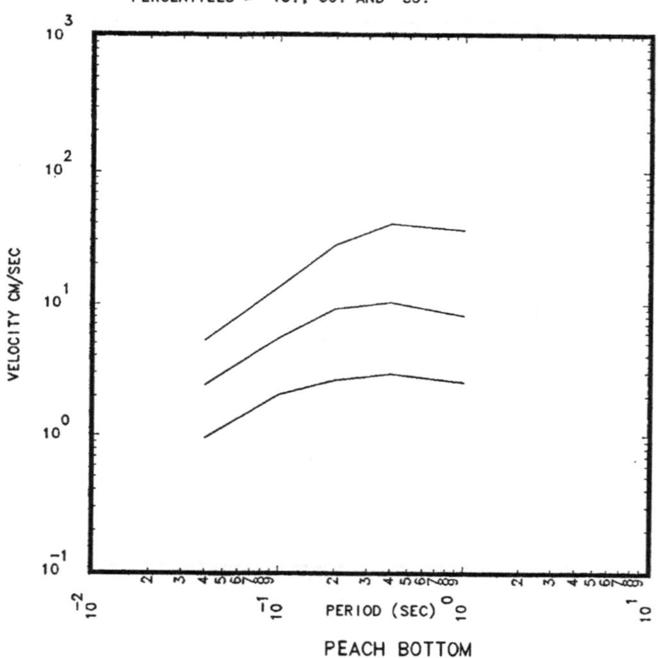

Figure 2.11.9 10000 year return period CPUHS for the 15th, 50th and 85th percentiles aggregated over all S and G-Experts for the Peach Bottom site.

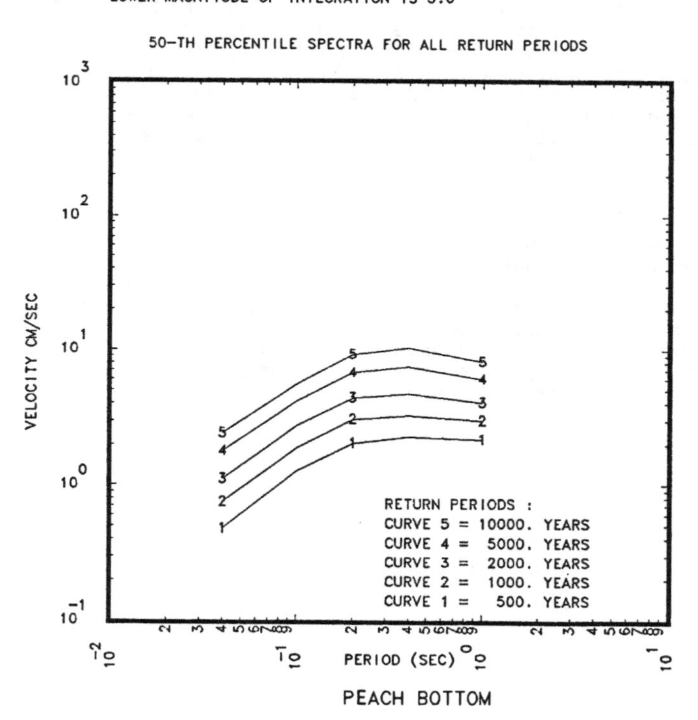

Figure 2.11.10 Comparison of the 50th percentile CPUHS for return periods of 500, 1000, 2000, 5000 and 10000 years for the Peach Bottom site.

2.12 PILGRIM

Pilgrim was placed in the sand-2 category (see Table 1.4) and is represented by the symbol "C" in Fig. 1.1. Table 2.12.1 and Figs. 2.12.1 to 2.12.10 give the basic results for the Pilgrim site. Interestingly, the BEHC is somewhat lower than the 50th percentile CPHC at higher PGA values. The AMHC is about the same as the 85th percentile CPHC at low PGA values and higher at high PGA. Figure 2.12.2 indicates that the diversity of opinion among the S-Experts is not particularly large, however from Fig. 2.12.3 we see a relatively large spread between the 15th and 85th percentile CPHCs. There is a relatively small spread between the G-Experts' BEHC per S-Expert thus the GM models are not the main source of the relatively wide spread between the 15th and 85th percentile CPHCs. Examination of the output for each S-Expert indicates that the relatively large spread between the 15th and 85th percentile curves comes from individual S-Experts uncertainties about the seismicity and zonation of the region around the Pilgrim site. In particular S-Experts 2,3,5,7 and 11 had wide uncertainty bounds (as compared to the other S-Experts) at the Pilgrim site.

We see from Table 2.12.1 that for most S-Experts the region around the site contributes most to the hazard. Only for S-Experts 7 and 13 do more distant zones become the most significant contributors to the hazard. As expected we see from Fig. 2.12.4 that smaller earthquakes contribute significantly to the hazard at the Pilgrim site. If earthquakes in the 3.75 to 5 magnitude range were included the PGA hazard would be increased well over a factor of 2 at 0.05g, about a factor of 2 at 0.3g and not dropping to about a factor of 1.0 until about 0.9g. The spread between the G-Experts BEHC per S-Expert are similar to that shown in Fig. 2.4.11.

TABLE 2.12.1

MOST IMPORTANT ZONES PER S-EXPERT
FOR PILGRIM

SITE SOIL CATEGORY SAND-2

S-XPT NUM.	HOST ZONE		ZONES CONTRIBUTING MOST SIGNIFICANTLY TO THE PGA BEHC ND % OF CONTRIBUTION AT LOW PGA(0.125G)				AT HIGH PGA(0.60G)			
1	ZONE 22	ZONE ID:	ZONE 22	ZONE 21	ZONE 20	ZONE 4	ZONE 22	ZONE 21	ZONE 20	ZONE 1
		% CONT.:	80.	11.	9.	0.	99.	1.	0.	0.
2	ZONE 31	ZONE ID:	ZONE 31	ZONE 32	COMP. ZON	ZONE 28	ZONE 31	ZONE 32	COMP. ZON	ZONE 27
		% CONT.:	78.	18.	4.	0.	93.	5.	2.	0.
3	ZONE 4	ZONE ID:	ZONE 4	ZONE 3	ZONE 2	COMP. ZON	ZONE 4	COMP. ZON	ZONE 3	ZONE 2
		% CONT.:	97.	2.	1.	0.	100.	0.	0.	0.
4	ZONE 23	ZONE ID:	ZONE 23	ZONE 18	ZONE 20	ZONE 16	ZONE 23	ZONE 18	ZONE 20	ZONE 16
		% CONT.:	50.	35.	8.	4.	84.	14.	2.	0.
5	ZONE 1	ZONE ID:	ZONE 1	ZONE 6	ZONE 3	ZONE 8	ZONE 1	ZONE 3	COMP. ZON	ZONE 8
		% CONT.:	93.	3.	3.	1.	100.	0.	0.	0.
6	ZONE 5	ZONE ID:	ZONE 5	ZONE 6	ZONE 3	ZONE 7	ZONE 5	ZONE 6	ZONE 3	ZONE 7
		% CONT.:	60.	30.	7.	2.	88.	11.	0.	0.
7	ZONE 24	ZONE ID:	ZONE 19	ZONE 24	ZONE 26	ZONE 17	ZONE 19	ZONE 24	ZONE 2 =	ZONE 17
		% CONT.:	90.	7.	1.	1.	95.	5.	0.	0.
10	ZONE 2	ZONE ID:	ZONE 2	ZONE 22	ZONE 23	ZONE 21	ZONE 2	ZONE 22	ZONE 19 =	ZONE 23
		% CONT.:	63.	24.	5.	3.	89.	7.	2.	1.
11	ZONE 1	ZONE ID:	ZONE 1	ZONE 3	ZONE 5	CZ = ZONE	ZONE 1	ZONE 3	CZ = ZONE	ZONE 5
		% CON T:	72.	23.	2.	1.	96.	3.	0.	0.
12	ZONE 32	ZONE ID:	ZONE 32	ZONE 33	ZONE 37	ZONE 34	ZONE 32	ZONE 37	ZONE 33	ZONE 34
		% CONT.:	80.	6.	6.	4.	99.	1.	0.	0.
13	CZ 15	ZONE ID:	ZONE 10	CZ 15	ZONE 12	ZONE 11	ZONE 10	CZ 15	ZONE 12	CZ 18
		% CONT.:	78.	15.	6.	1.	53.	47.	0.	0.

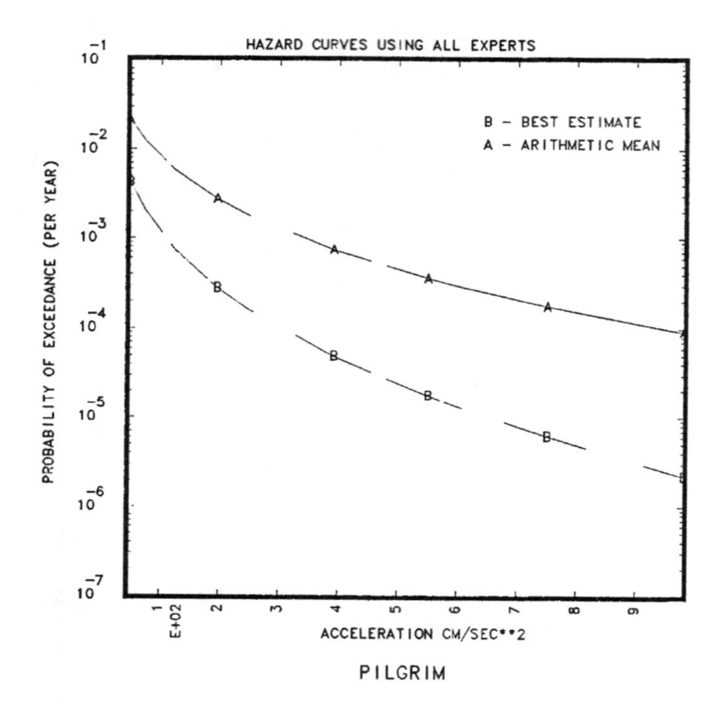

Figure 2.12.1 Comparison of the BEHC and AMHC aggregated over all S and G-Experts for the Pilgrim site.

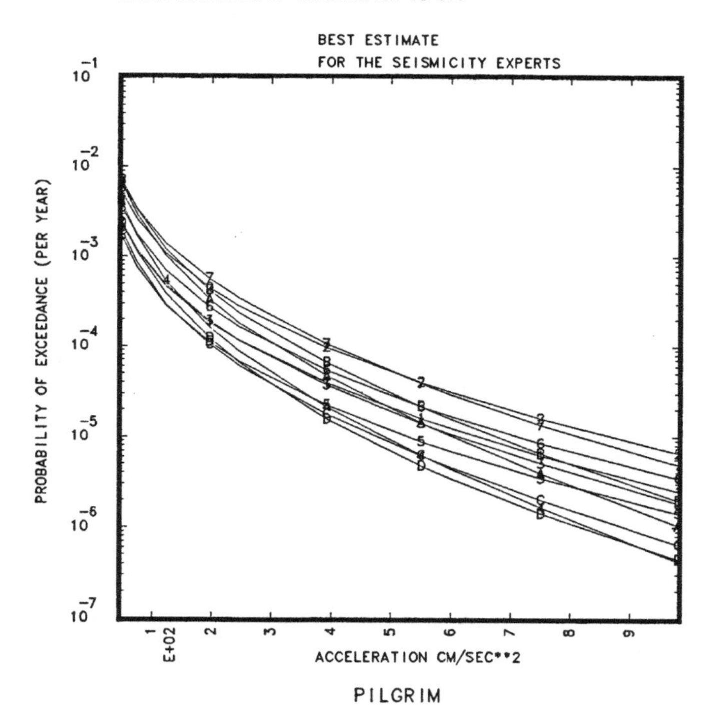

Figure 2.12.2 BEHCs per S-Expert combined over all G-Experts for the Pilgrim site. Plot symbols given in Table 2.0.

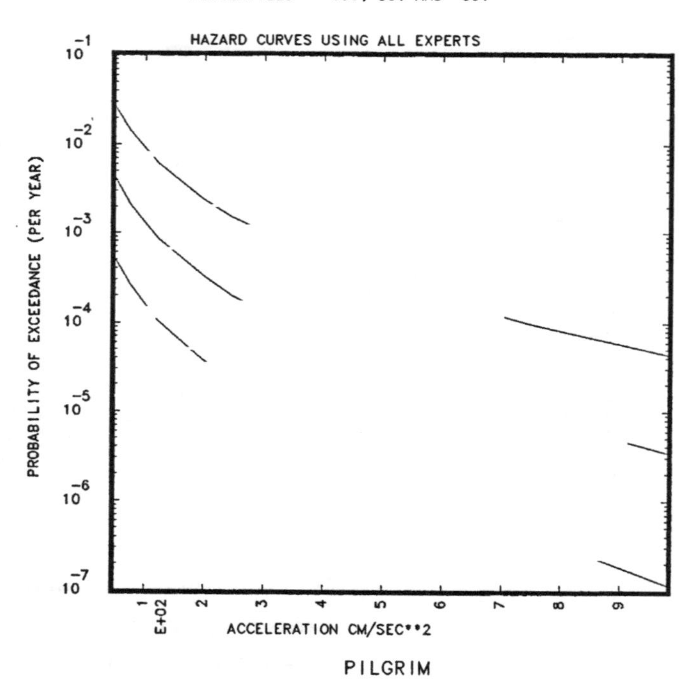

Figure 2.12.3 CPHCs for the 15th, 50th and 85th percentiles based on all S and G-Experts' input for the Pilgrim site.

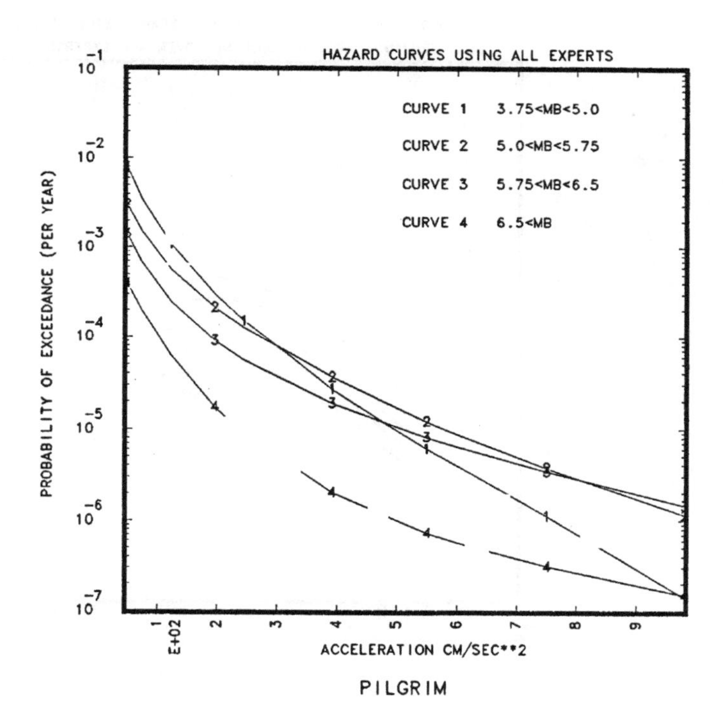

Figure 2.12.4 BEHCs which include only the contribution to the PGA hazard from earthquakes within the indicated magnitude range for the Pilgrim site.

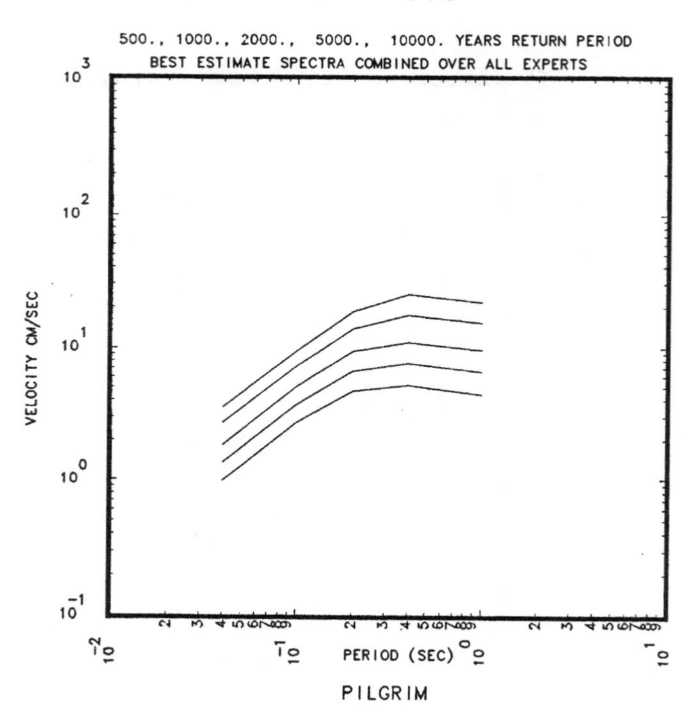

Figure 2.12.5 BEUHS for return periods of 500, 1000, 2000, 5000 and 10000 years aggregated over all S and G-Experts for the Pilgrim site.

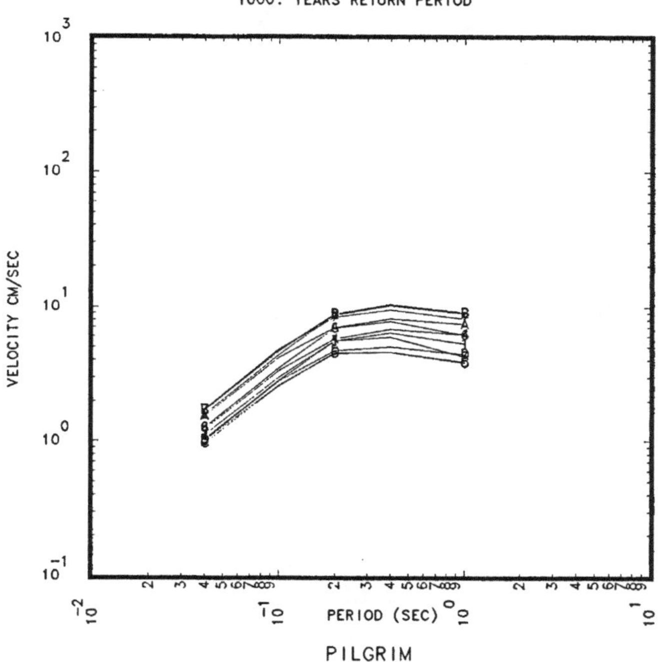

Figure 2.12.6 The 1000 year return period BEUHS per S-Expert aggregated over all G-Experts for the Pilgrim site. Plot symbols are given in Table 2.0.

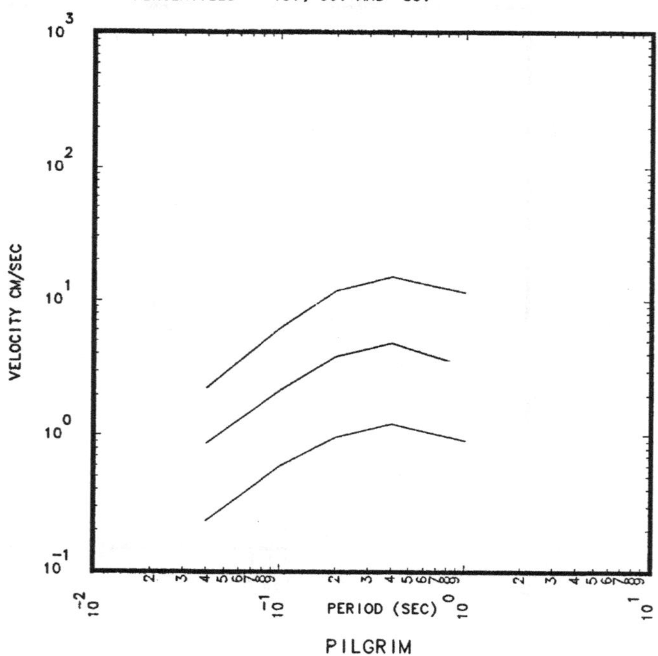

Figure 2.12.7 500 year return period CPUHS for the 15th, 50th and 85th percentiles aggregated over all S and G-Experts for the Pilgrim site.

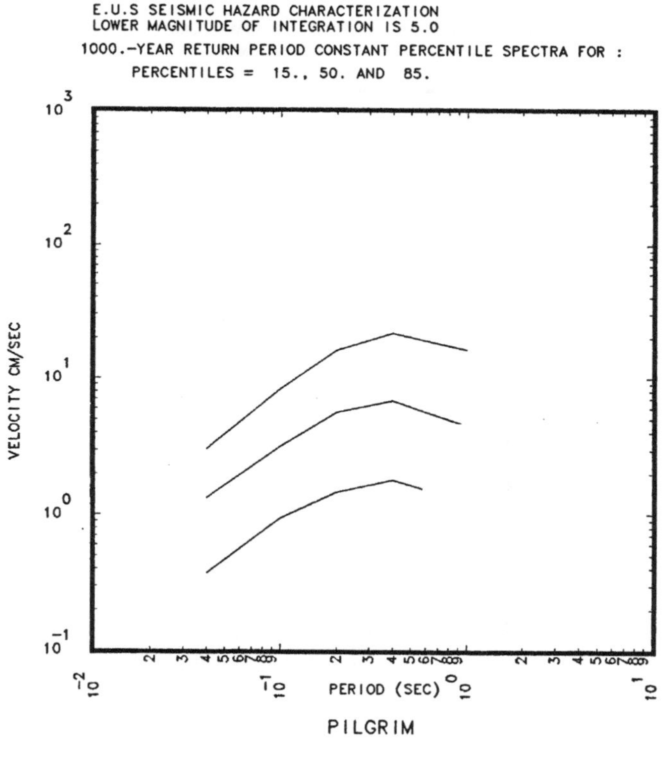

E.U.S SEISMIC HAZARD CHARACTERIZATION
LOWER MAGNITUDE OF INTEGRATION IS 5.0
1000.-YEAR RETURN PERIOD CONSTANT PERCENTILE SPECTRA FOR :
PERCENTILES = 15., 50. AND 85.

VELOCITY CM/SEC

PERIOD (SEC)

PILGRIM

Figure 2.12.8 1000 year return period CPUHS for the 15th, 50th and 85th percentile aggregated over all S and G-Experts for the Pilgrim site.

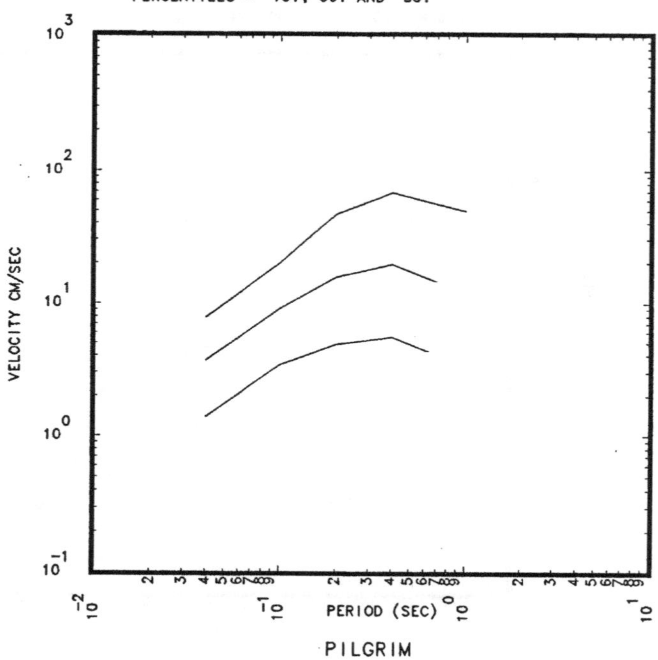

Figure 2.12.9 10000 year return period CPUHS for the 15th, 50th and 85th percentiles aggregated over all S and G-Experts for the Pilgrim site.

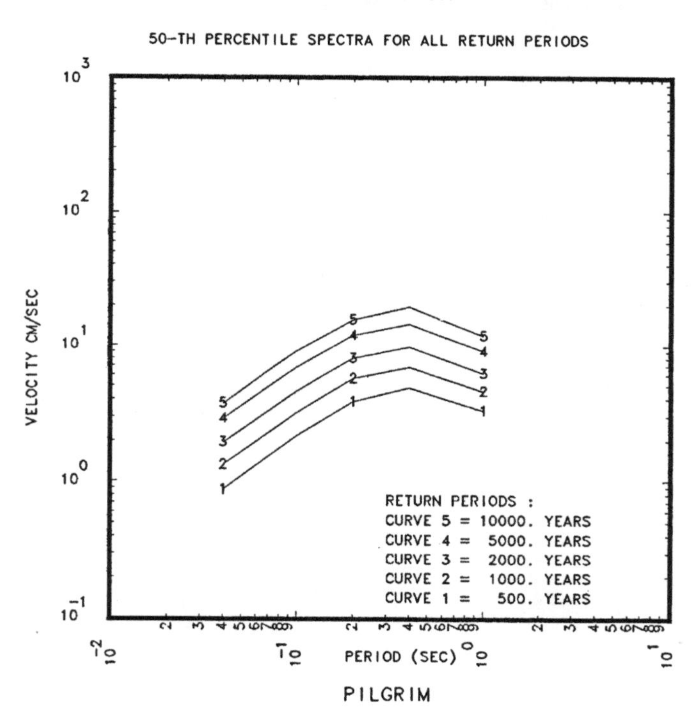

Figure 2.12.10 Comparison of the 50th percentile CPUHS for return periods of 500, 1000, 2000, 5000 and 10000 years for the Pilgrim site.

2.13 SALEM

The Salem site is a deep soil site and is represented by the symbol "D" in Fig. 1.1. Table 2.13.1 and Figs. 2.13.1 to 2.13.10 give the basic results for the Salem site. We see from Fig. 1.1 that Salem is located next to the Hope Creek site. The results are essentially the same as for the Hope Creek site, hence the discussion of Section 2.4 will not be repeated. The difference between the results in Section 2.4 and this section are due to the differences in the random sampling of the distributions in the Monte Carlo analysis and the fact that the uncertainty in the zonation seismicity parameter and ground motion models is large.

TABLE 2.13.1

MOST IMPORTANT ZONES PER S-EXPERT
FOR SALEM

SITE SOIL CATEGORY DEEP-SOIL

S-XPT NUM.	HOST ZONE		ZONES CONTRIBUTING MOST SIGNIFICANTLY TO THE PGA BEHC AND % OF CONTRIBUTION AT LOW PGA(0.125G)				AT HIGH PGA(0.60G)			
1	ZONE 1	ZONE ID:	ZONE 1	ZONE 4	ZONE 20	ZONE 21	ZONE 1	ZONE 4	ZONE 3	ZONE 2
		% CONT.:	68.	26.	3.	1.	95.	5.	0.	0.
2	ZONE 28	ZONE ID:	ZONE 28	ZONE 30	ZONE 32	ZONE 27	ZONE 28	COMP. ZON	ZONE 32	ZONE 30
		% CON T:	95.	2.	1.	1.	99.	0.	0.	0.
3	ZONE 8A	ZONE ID:	ZONE 8A	ZONE 5	ZONE 4	ZONE 2	ZONE 8A	ZONE 5	COMP. ZON	ZONE 3
		% CON T:	57.	41.	1.	0.	86.	14.	0.	0.
4	COMP. ZO	ZONE ID:	ZONE 11	ZONE 12	COMP. ZON	ZONE 16	COMP. ZON	ZONE 11	ZONE 8	ZONE 7
		% CONT.:	36.	15.	14.	13.	68.	30.	1.	0.
5	ZONE 8	ZONE ID:	ZONE 1	ZONE 8	ZONE 9	ZONE 6	ZONE 1	ZONE 8	COMP. ZON	ZONE 3
		% CONT.:	57.	27.	8.	7.	73.	27.	0.	0.
6	ZONE 6	ZONE ID:	ZONE 6	ZONE 13	ZONE 7	ZONE 15	ZONE 6	COMP. ZON	ZONE 7	ZONE 2
		% CON T:	98.	1.	1.	0.	100.	0.	0.	0.
7	ZONE 29	ZONE ID:	ZONE 29	ZONE 7	ZONE 13	ZONE 14	ZONE 29	ZONE 7	ZONE 2 =	ZONE 13
		% CONT.:	85.	11.	2.	1.	95.	4.	0.	0.
10	ZONE 4B	ZONE ID:	ZONE 4B	ZONE 5	ZONE 19 =	ZONE 6	ZONE 4B	ZONE 19 =	ZONE 1	ZONE
		% CONT.:	96.	2.	1.	0.	98.	2.	0.	0.
11	CZ = ZON	ZONE ID:	ZONE 5	CZ = ZONE	ZONE 8	ZONE 3	ZONE 5	CZ = ZONE	ZONE 8	ZONE 1
		% CONT.:	61.	33.	2.	2.	60.	39.	0.	0.
12	ZONE 32	ZONE ID:	ZONE 32	ZONE 31	ZONE 23A	ZONE 27	ZONE 32	ZONE 17	ZONE 19	ZONE 20
		% CONT.:	99.	0.	0.	0.	100.	0.	0.	0.
13	CZ 17	ZONE ID:	CZ 17	CZ 15	ZONE 10	ZONE 9	CZ 17	CZ 15	ZONE 7	ZONE 8
		% CONT.:	92.	5.	2.	0.	98.	2.	0.	0.

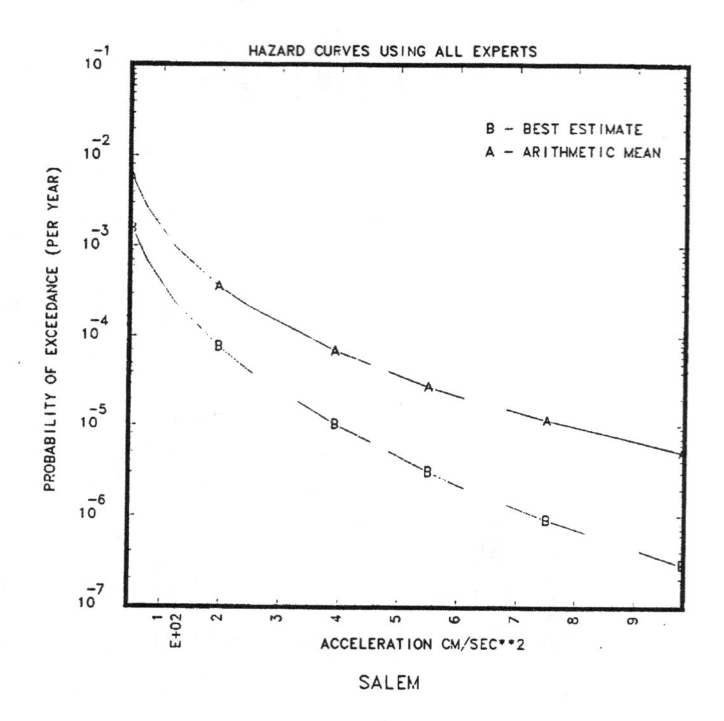

Figure 2.13.1 Comparison of the BEHC and AMHC aggregated over all S and G-Experts for the Salem site.

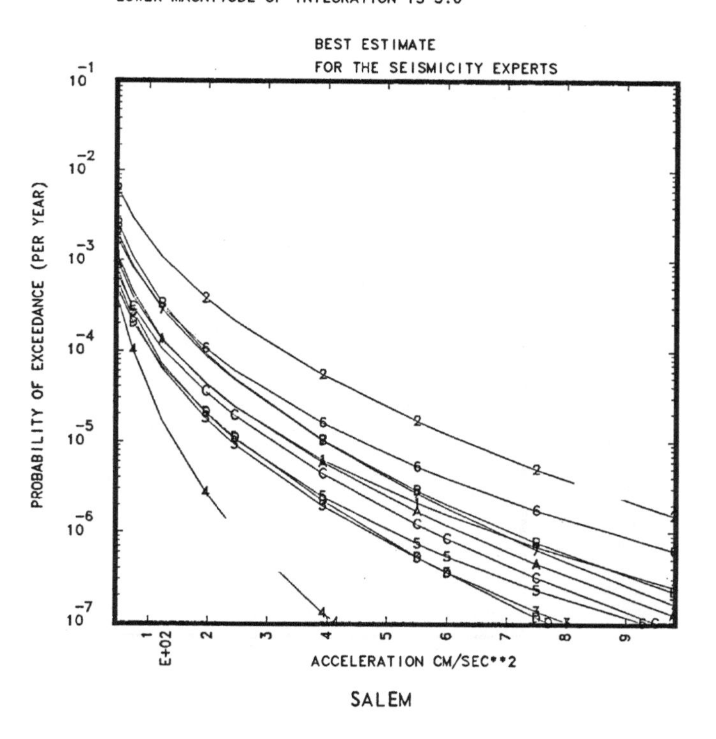

Figure 2.13.2 BEHCs per S-Expert combined over all G-Experts for the Salem site. Plot symbols given in Table 2.0.

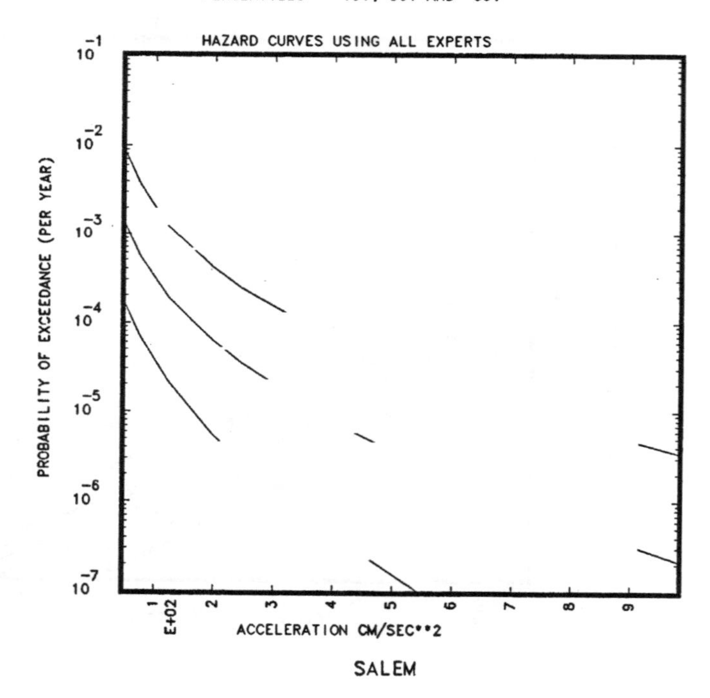

Figure 2.13.3 CPHCs for the 15th, 50th and 85th percentiles based on all S and G-Experts' input for the Salem site.

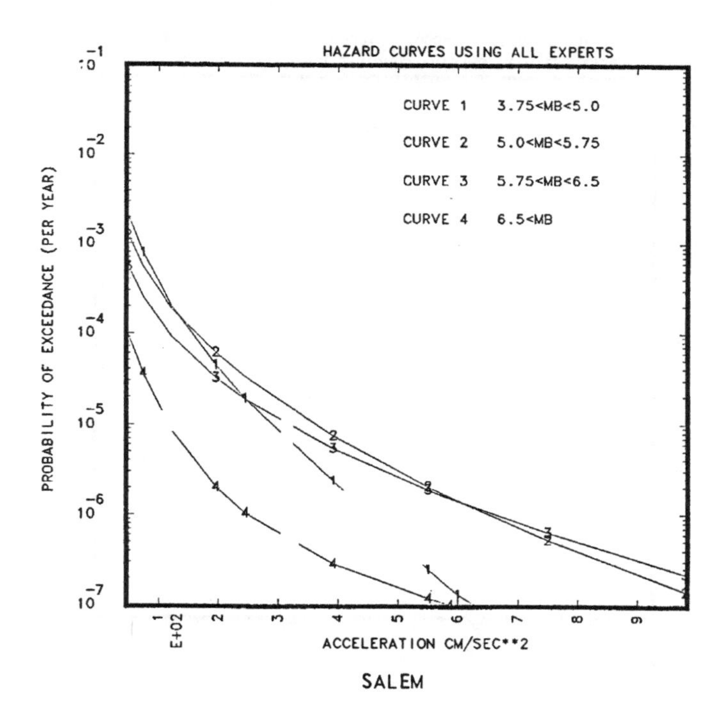

Figure 2.13.4 BEHCs which include only the contribution to the PGA hazard from earthquakes within the indicated magnitude range for the Salem site.

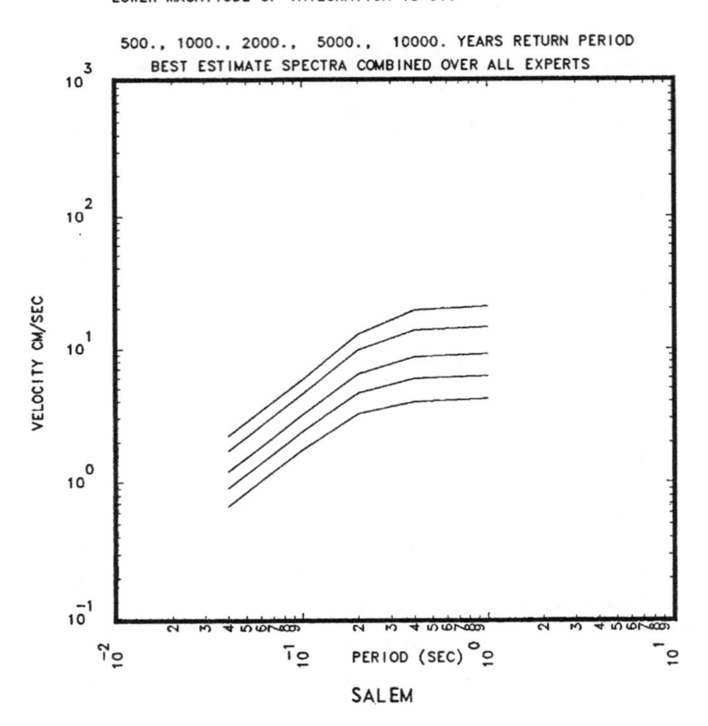

Figure 2.13.5 BEUHS for return periods of 500, 1000, 2000, 5000 and 10000 years aggregated over all S and G-Experts for the Salem site.

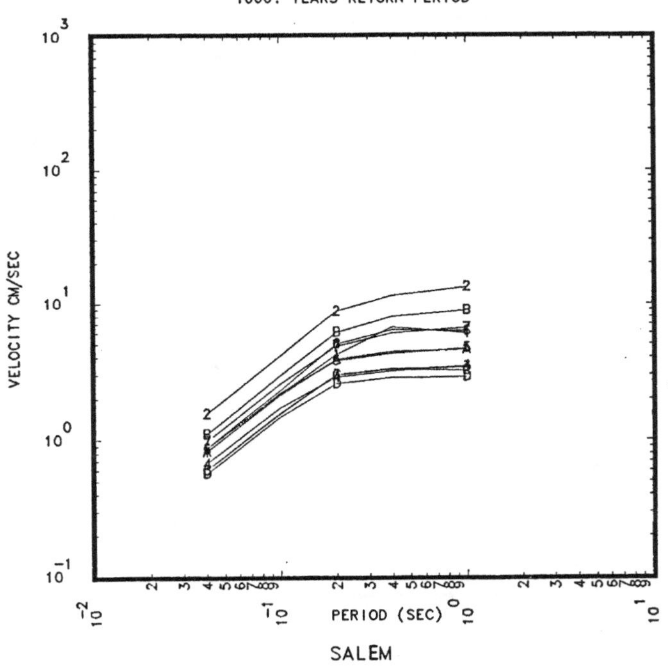

Figure 2.13.6 The 1000 year return period BEUHS per S-Expert aggregated over all G-Experts for the Salem site. Plot symbols are given in Table 2.0.

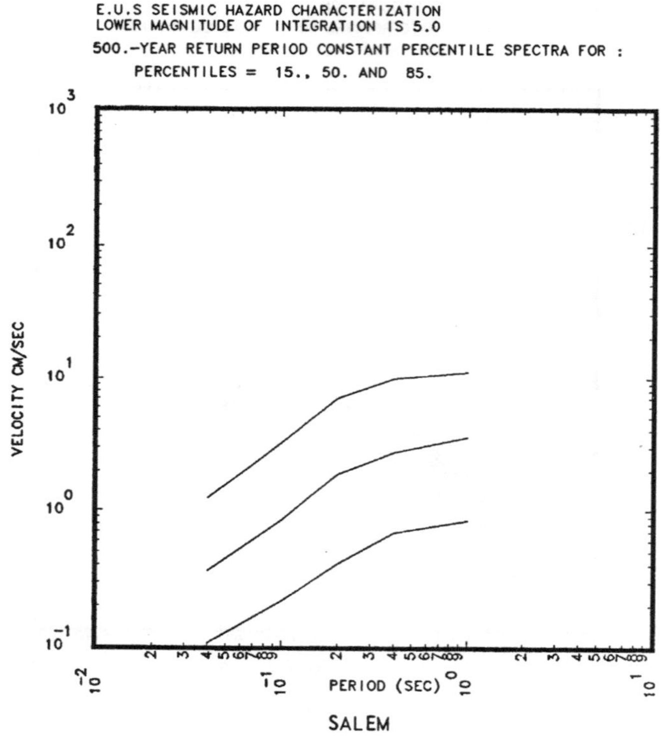

E.U.S SEISMIC HAZARD CHARACTERIZATION
LOWER MAGNITUDE OF INTEGRATION IS 5.0
500.-YEAR RETURN PERIOD CONSTANT PERCENTILE SPECTRA FOR :
PERCENTILES = 15., 50. AND 85.

VELOCITY CM/SEC

PERIOD (SEC)

SALEM

Figure 2.13.7 500 year return period CPUHS for the 15th, 50th and 85th percentiles aggregated over all S and G-Experts for the Salem site.

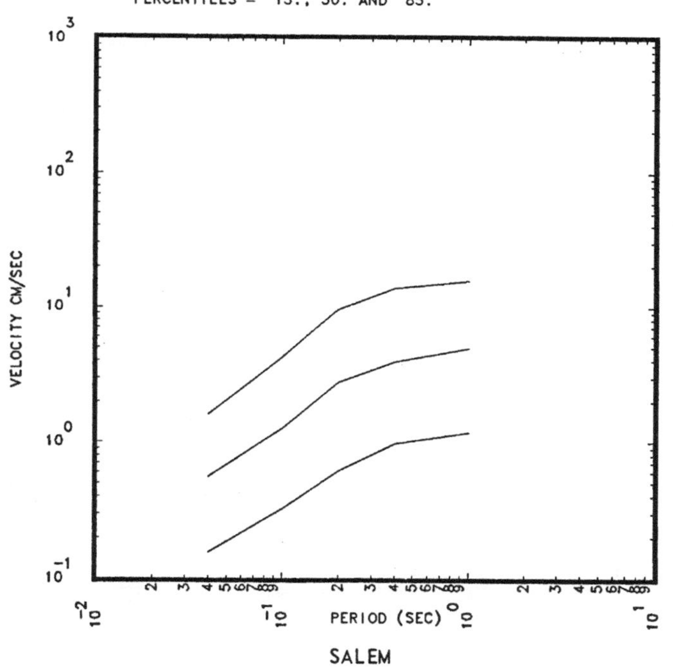

Figure 2.13.8 1000 year return period CPUHS for the 15th, 50th and 85th percentile aggregated over all S and G-Experts for the Salem site.

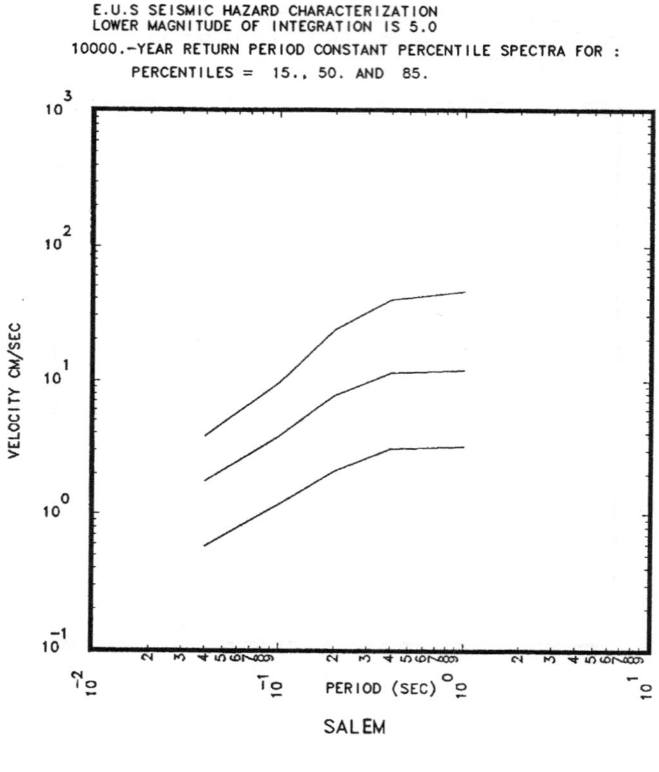

E.U.S SEISMIC HAZARD CHARACTERIZATION
LOWER MAGNITUDE OF INTEGRATION IS 5.0

10000.-YEAR RETURN PERIOD CONSTANT PERCENTILE SPECTRA FOR :

PERCENTILES = 15., 50. AND 85.

VELOCITY CM/SEC

PERIOD (SEC)

SALEM

Figure 2.13.9 10000 year return period CPUHS for the 15th, 50th and 85th percentiles aggregated over all S and G-Experts for the Salem site.

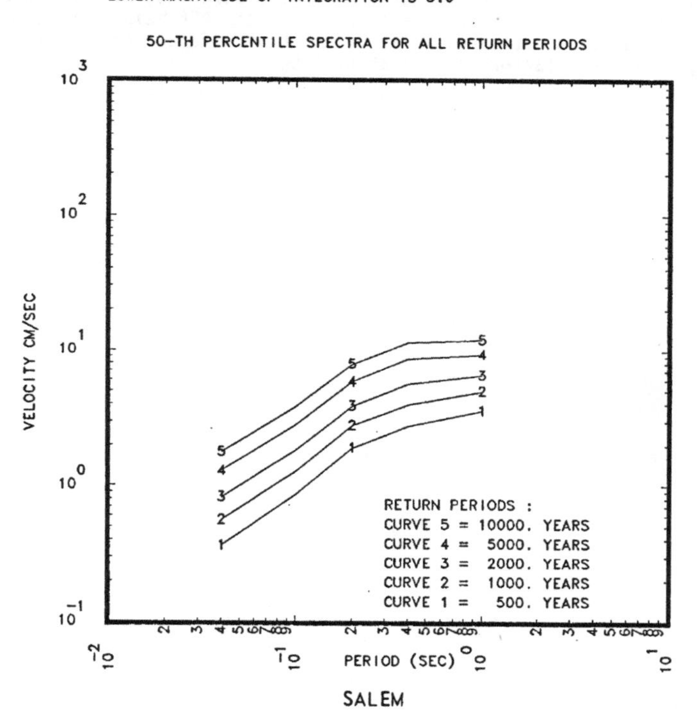

Figure 2.13.10 Comparison of the 50th percentile CPUHS for return periods of 500, 1000, 2000, 5000 and 10000 years for the Salem site.

2.14 SEABROOK

The Seabrook site is a rock site and is represented by the symbol "E" in Fig. 1.1. Table 2.14.1 and Figs. 2.14.1 to 2.14.10 give the basic results for the Seabrook site. The AMHC is about the same as the 85th percentile CPHC.

We see from Table 2.14 that for most S-Experts the zone that contains the site contributes most to the hazard. Thus, as expected, the spread between the G-Experts' BEHCs per S-Expert is typical for rock sites and is similar to that shown in Fig. 2.1.11. Also, we see from Fig. 2.14.4 that if earthquakes in the 3.75 to 5.0 magnitude range were included, the PGA hazard curve would be increased by about a factor of 2. at 0.05g about a factor of 1.3 at 0.15g and dropping to about a factor of 1 at about 0.5g.

TABLE 2.14.1

MOST IMPORTANT ZONES PER S-EXPERT
FOR SEABROOK

SITE SOIL CATEGORY ROCK

S-XPT NUM.	HOST ZONE		AT LOW PGA(0.125G)				AT HIGH PGA(0.60G)			
			ZONES CONTRIBUTING MOST SIGNIFICANTLY TO THE PGA BEHC AND % OF CONTRIBUTION							
1	ZONE 22	ZONE ID:	ZONE 22	ZONE 20	ZONE 21	ZONE 4	ZONE 22	ZONE 21	ZONE 20	ZONE 1
		% CONT.:	57.	22.	21.	1.	83.	10.	6.	0.
2	ZONE 31	ZONE ID:	ZONE 31	ZONE 32	COMP. ZON	ZONE 28	ZONE 31	ZONE 32	COMP. ZON	ZONE 27
		% CONT.:	62.	36.	1.	1.	69.	31.	0.	0.
3	ZONE 4	ZONE ID:	ZONE 4	ZONE 3	ZONE 2	ZONE 5	ZONE 4	ZONE 3	ZONE 2	COMP. ZON
		% CONT.:	83.	11.	5.	0.	98.	1.	0.	0.
4	ZONE 18	ZONE ID:	ZONE 18	ZONE 20	ZONE 16	ZONE 19	ZONE 18	ZONE 20	ZONE 16	ZONE 19
		% CONT.:	83.	8.	4.	4.	96.	4.	0.	0.
5	ZONE 1	ZONE ID:	ZONE 1	ZONE 6	ZONE 3	ZONE 4	ZONE 1	ZONE 3	ZONE 4	[illegible]. ZON
		% CONT.:	75.	12.	10.	2.	96.	4.	0.	0.
6	ZONE 5	ZONE ID:	ZONE 5	ZONE 3	ZONE 7	ZONE 6	ZONE 5	ZONE 3	ZONE 6	ZONE 7
		% CON T:	64.	17.	10.	8.	92.	3.	2.	2.
7	ZONE 19	ZONE ID:	ZONE 19	ZONE 17	ZONE 26	ZONE 18	ZONE 19	ZONE 26	ZONE 17	ZONE 24
		% CONT.:	89.	4.	3.	2.	99.	0.	0.	0.
10	ZONE 23	ZONE ID:	ZONE 23	ZONE 22	ZONE 21	ZONE 2	ZONE 23	ZONE 22	ZONE 21	ZONE 2
		% CON T:	42.	28.	16.	3.	57.	34.	8.	0.
11	ZONE 1	ZONE ID:	ZONE 3	ZONE 1	ZONE 5	ZONE 2	ZONE 1	ZONE 3	ZONE 5	ZONE 4
		% CONT.:	54.	36.	3.	3.	53.	45.	1.	1.
12	ZONE 33	ZONE ID:	ZONE 33	ZONE 32	ZONE 31	ZONE 34	ZONE 33	ZONE 32	ZONE 34	ZONE 31
		% CONT.:	66.	15.	9.	9.	92.	6.	1.	0.
13	ZONE 10	ZONE ID:	ZONE 10	ZONE 12	ZONE 11	CZ 15	ZONE 10	ZONE 12	CZ 15	ZONE 11
		% CON T:	82.	12.	3.	3.	97.	2.	1.	0.

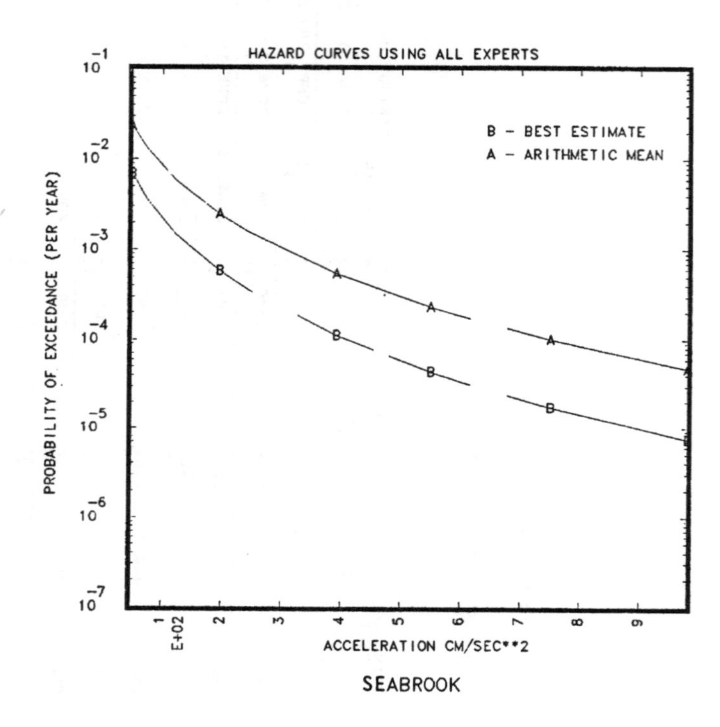

Figure 2.14.1 Comparison of the BEHC and AMHC aggregated over all S and G-Experts for the Seabrook site.

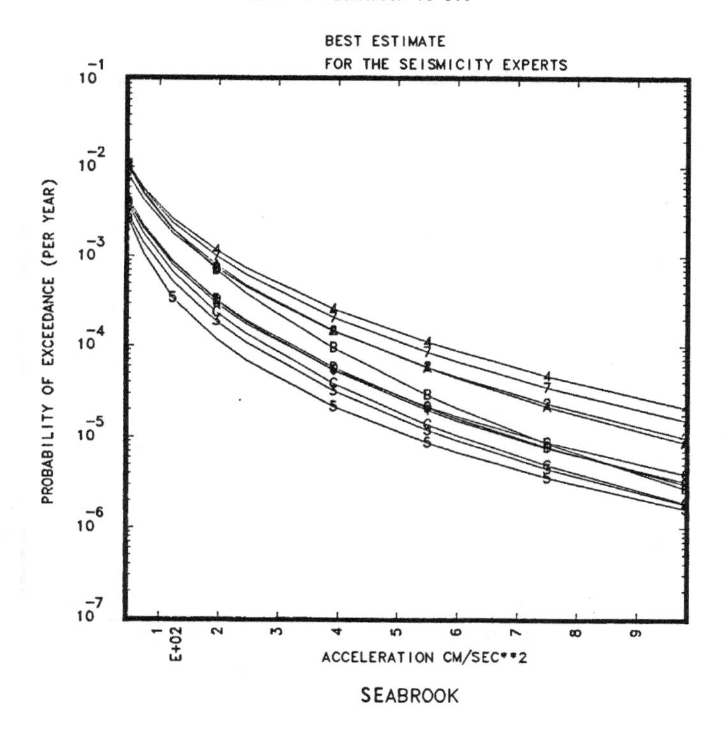

Figure 2.14.2 BEHCs per S-Expert combined over all G-Experts for the Seabrook site. Plot symbols given in Table 2.0.

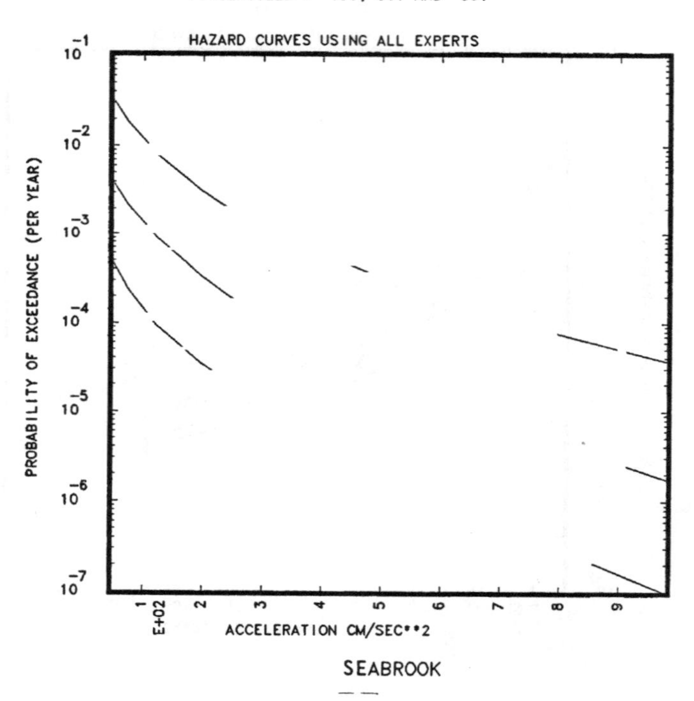

Figure 2.14.3 CPHCs for the 15th, 50th and 85th percentiles based on all S and G-Experts' input for the Seabrook site.

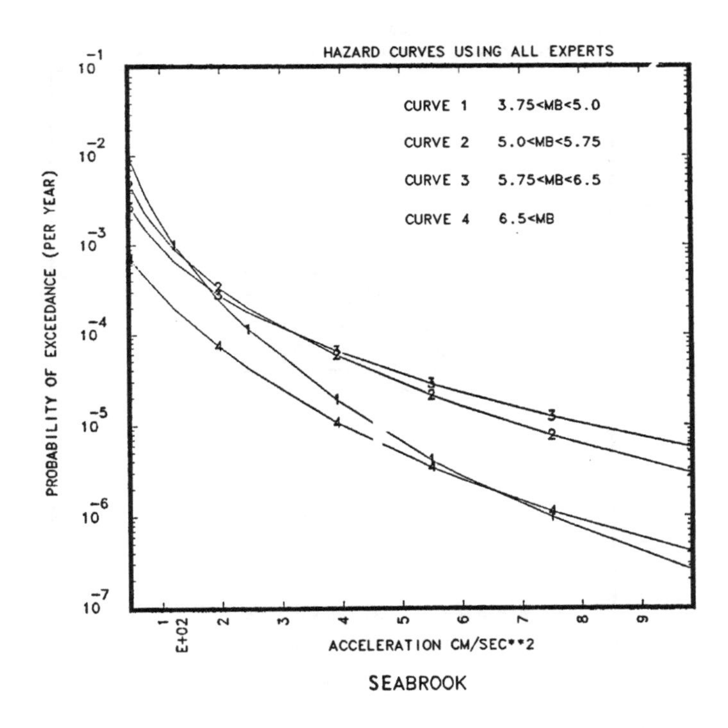

Figure 2.14.4 BEHCs which include only the contribution to the PGA hazard from earthquakes within the indicated magnitude range for the Seabrook site.

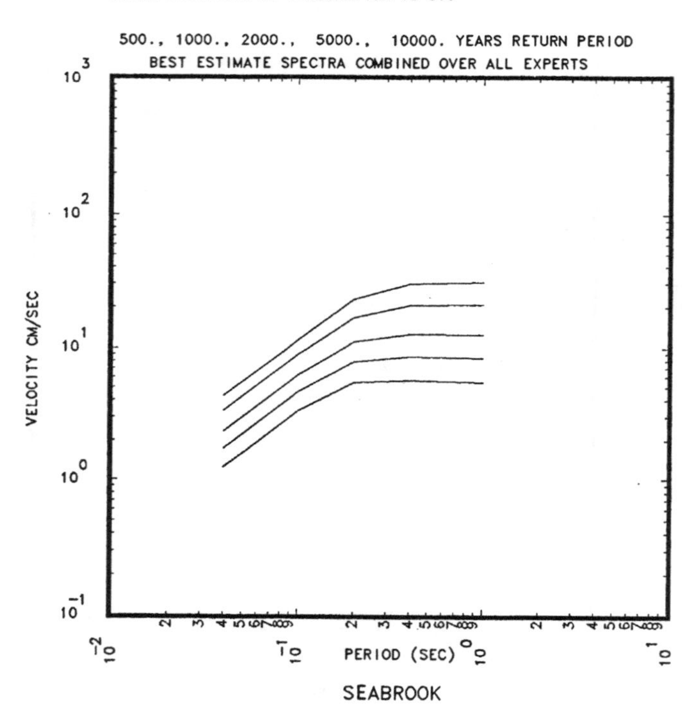

Figure 2.14.5 BEUHS for return periods of 500, 1000, 2000, 5000 and 10000 years aggregated over all S and G-Experts for the Seabrook site.

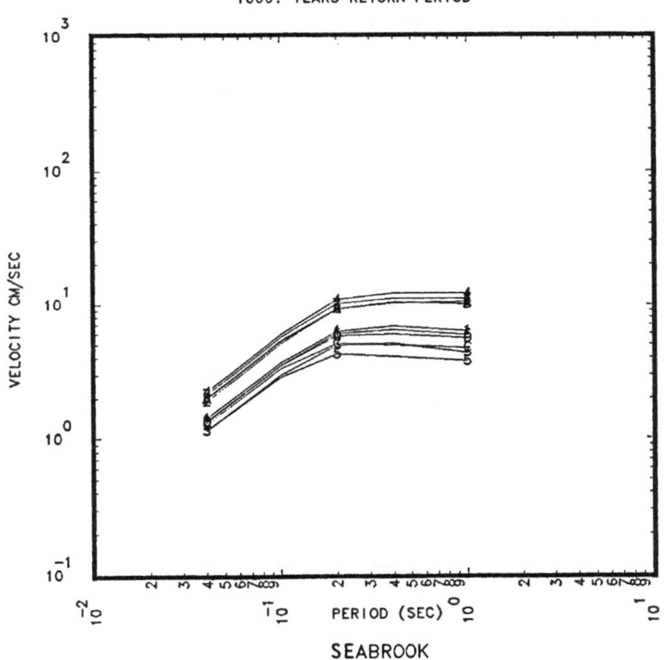

Figure 2.14.6 The 1000 year return period BEUHS per S-Expert aggregated over all G-Experts for the Seabrook site. Plot symbols are given in Table 2.0.

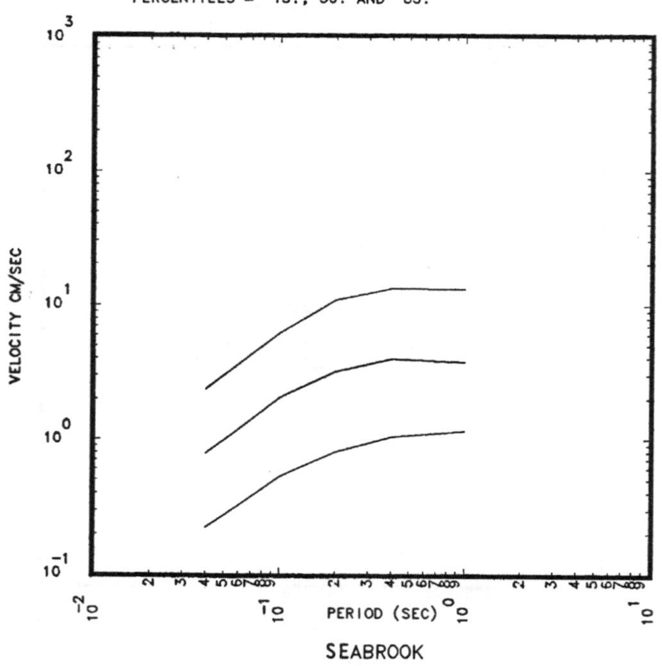

Figure 2.14.7 500 year return period CPUHS for the 15th, 50th and 85th percentiles aggregated over all S and G-Experts for the Seabrook site.

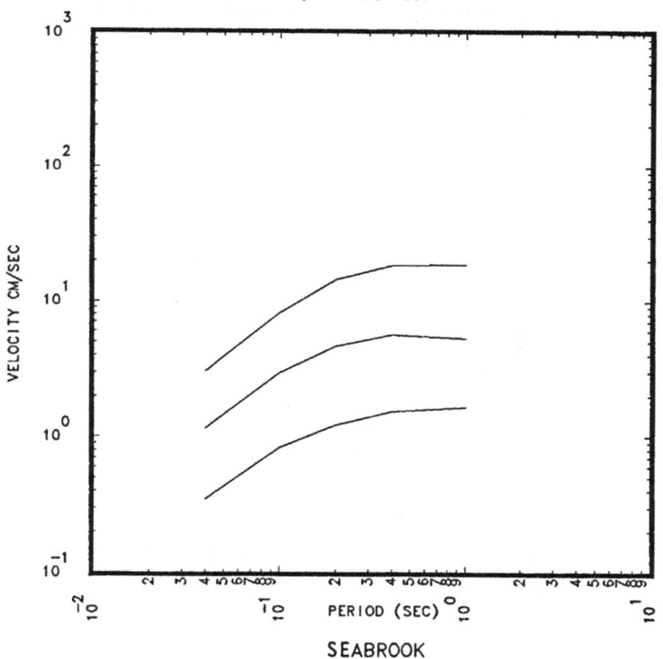

Figure 2.14.8 1000 year return period CPUHS for the 15th, 50th and 85th percentile aggregated over all S and G-Experts for the Seabrook site.

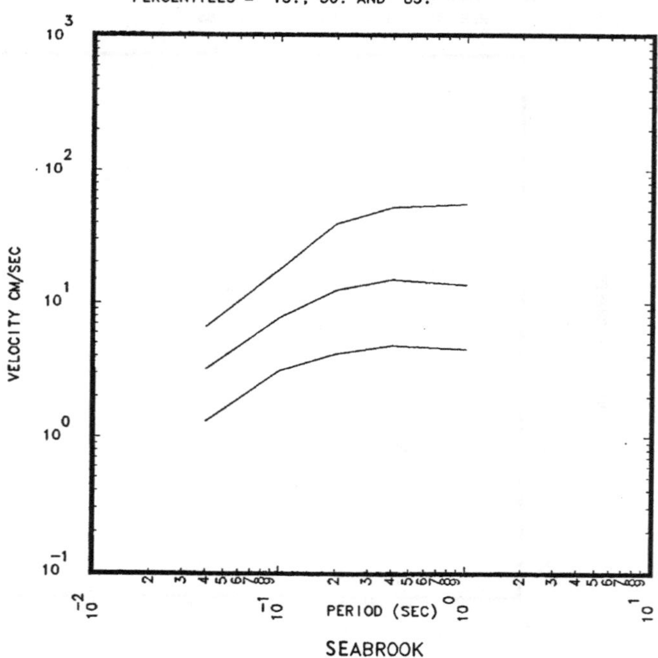

Figure 2.14.9 10000 year return period CPUHS for the 15th, 50th and 85th percentiles aggregated over all S and G-Experts for the Seabrook site.

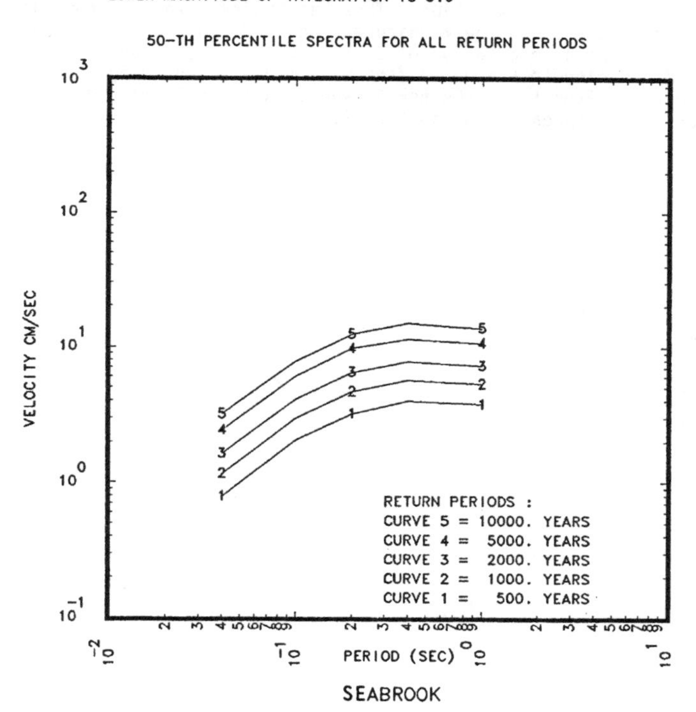

Figure 2.14.10 Comparison of the 50th percentile CPUHS for return periods of 500, 1000, 2000, 5000 and 10000 years for the Seabrook site.

2.15 SHOREHAM

The Shoreham site is a deep soil site and is represented by the symbol "F" in Fig. 1.1. Table 2.15.1 and Figs. 2.15.1 to 2.15.10 give the basic results for the Shoreham site. Interestingly, the BEHC is slightly below the 50th percentile CPHC at higher PGA values.

We see from Table 2.15.1 that, for all but one S-Expert, the zone which contains the site is also the zone which contributes most to the BEHC. S-Expert 4 is the only exception. Thus, as typical, we see from Fig. 2.15.4 that if earthquakes in the magnitude range 3.75 to 5.0 were included the PGA hazard would be over a factor of 2 at 0.05g higher, a factor of about 1.3 at 0.3g and dropping to a about a factor of 1.0 at about 0.6g. The spread between the G-Experts' BEHCs per S-Expert is typical for a deep soil site and similar to the spread shown in Fig. 2.4.11.

TABLE 2.15.1

MOST IMPORTANT ZONES PER S-EXPERT FOR SHOREHAM

SITE SOIL CATEGORY DEEP-SOIL

S-XPT N.	HOST ZONE		ZONES CONTRIBUTING MOST SIGNIFICANTLY TO THE PGA BEHC AND % OF CONTRIBUTION AT LOW PGA(0.125G)				AT HIGH PGA(0.60G)			
1	ZONE 1	ZONE ID:	ZONE 1	ZONE 22	ZONE 20	ZONE 21	ZONE 1	ZONE 22	ZONE 4	ZONE 2
		% CONT.:	40.	35.	13.	7.	79.	20.	1.	1.
2	ZONE 31	ZONE ID:	ZONE 31	ZONE 32	ZONE 28	COMP. ZON	ZONE 31	ZONE 32	COMP. ZON	ZONE 2
		% CONT.:	82.	10.	5.	2.	97.	2.	2.	0.
3	ZONE 4	ZONE ID:	ZONE 4	ZONE 5	ZONE 2	ZONE 3	ZONE 4	COMP. ZON	ZONE 5	ZONE 2
		% CONT.:	97.	2.	1.	1.	100.	0.	0.	0.
4	COMP. ZO	ZONE ID:	ZONE 11	ZONE 23	ZONE 18	ZONE 16	ZONE 11	COMP. ZON	ZONE 23	ZONE 1
		% CONT.:	31.	18.	17.	17.	59.	30.	10.	1.
5	ZONE 1	ZONE ID:	ZONE 1	ZONE 3	ZONE 6	ZONE 8	ZONE 1	ZONE 8	COMP. ZON	ZONE
		% CONT.:	96.	1.	1.	1.	100.	0.	0.	0.
6	ZONE 6	ZONE ID:	ZONE 6	ZONE 7	ZONE 5	ZONE 3	ZONE 6	ZONE 7	COMP. ZON	ZONE 5
		% CONT.:	94.	3.	1.	1.	99.	0.	0.	0.
7	ZONE 29	ZONE ID:	ZONE 29	ZONE 14	ZONE 24	ZONE 19	ZONE 29	ZONE 24	ZONE 14	ZONE 2
		% CONT.:	49.	27.	12.	3.	81.	10.	9.	0.
10	ZONE 4A	ZONE ID:	ZONE 4A	ZONE 4B	ZONE 2	ZONE 19 =	ZONE 4A	ZONE 4B	ZONE 19 =	ZONE
		% CONT.:	71.	15.	3.	3.	93.	5.	1.	0.
11	CZ = ZON	ZONE ID:	ZONE 5	CZ = ZONE	ZONE 3	ZONE 1	CZ = ZONE	ZONE 5	ZONE 1	ZONE
		% CONT.:	41.	32.	13.	9.	50.	47.	2.	1.
12	ZONE 32	ZONE ID:	ZONE 32	ZONE 34	ZONE 31	ZONE 33	ZONE 32	ZONE 25	ZONE 19	ZONE 2
		% CONT.:	97.	1.	1.	0.	100.	0.	0.	0.
13	ZONE 10	ZONE ID:	ZONE 10	CZ 17	ZONE 12	CZ 15	ZONE 10	CZ 15	CZ 17	ZONE
		% CONT.:	97.	1.	1.	1.	100.	0.	0.	0.

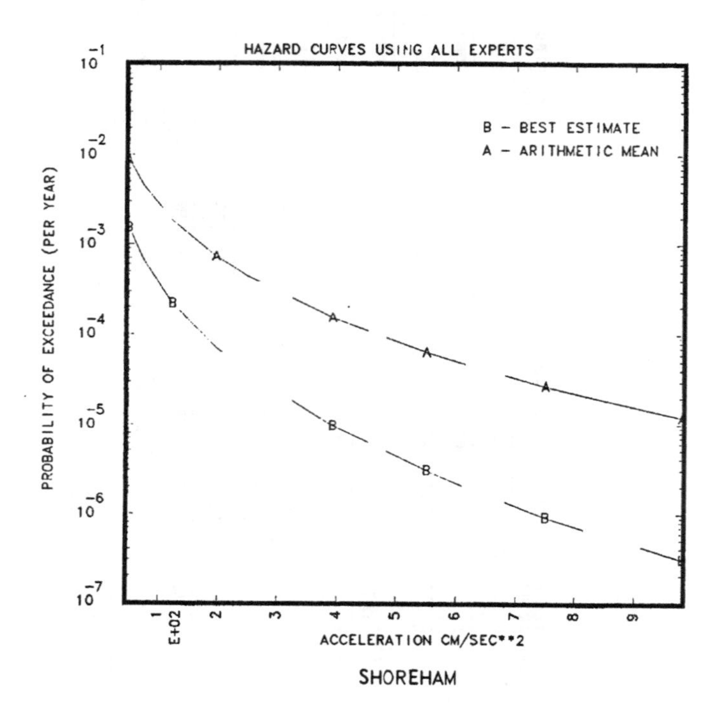

Figure 2.15.1 Comparison of the BEHC and AMHC aggregated over all S and G-Experts for the Shoreham site.

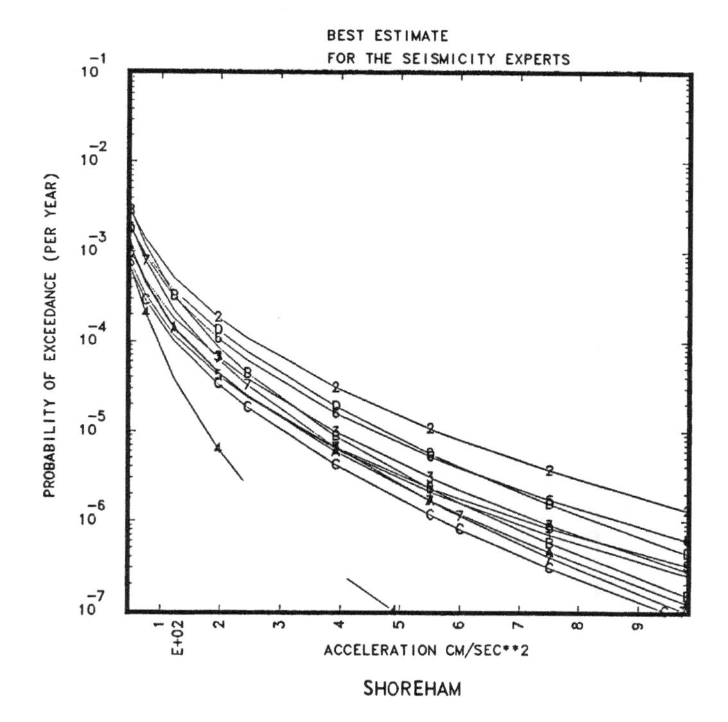

Figure 2.15.2 BEHCs per S-Expert combined over all G-Experts for the Shoreham site. Plot symbols given in Table 2.0.

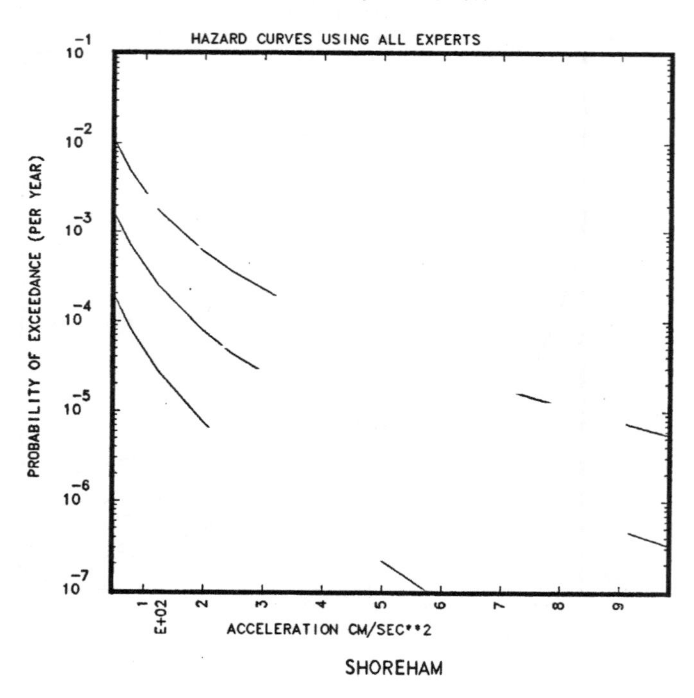

Figure 2.15.3 CPHCs for the 15th, 50th and 85th percentiles based on all S and G-Experts' input for the Shoreham site.

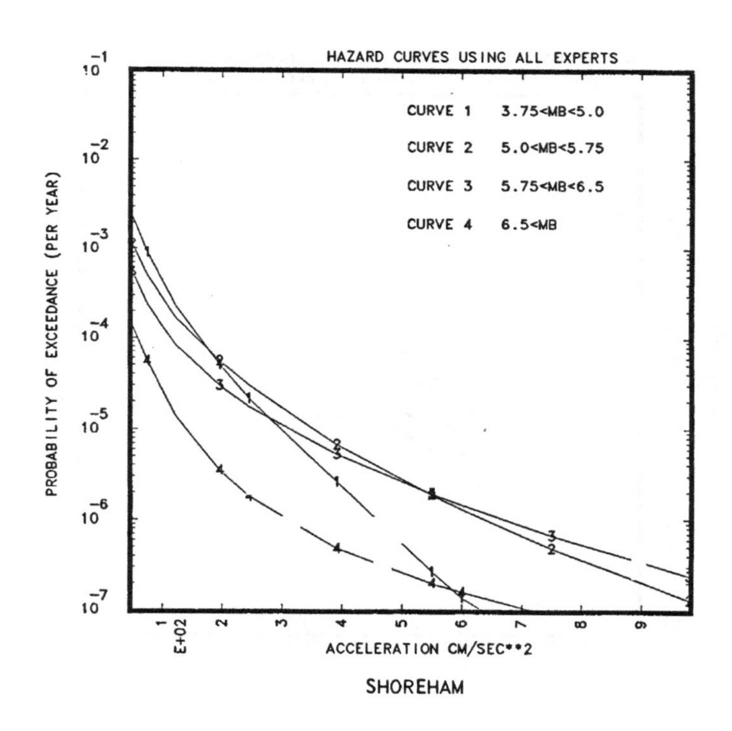

Figure 2.15.4 BEHCs which include only the contribution to the PGA hazard from earthquakes within the indicated magnitude range for the Shoreham site.

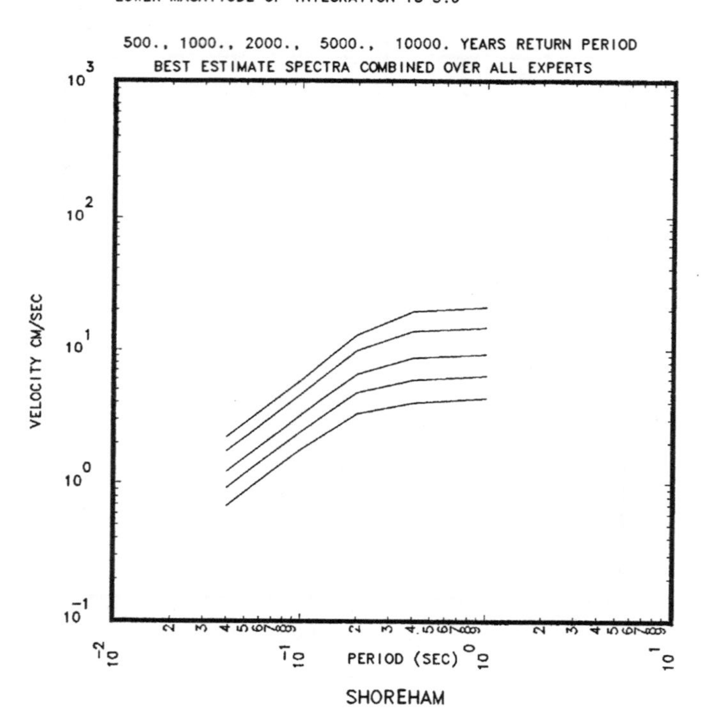

Figure 2.15.5 BEUHS for return periods of 500, 1000, 2000, 5000 and 10000 years aggregated over all S and G-Experts for the Shoreham site.

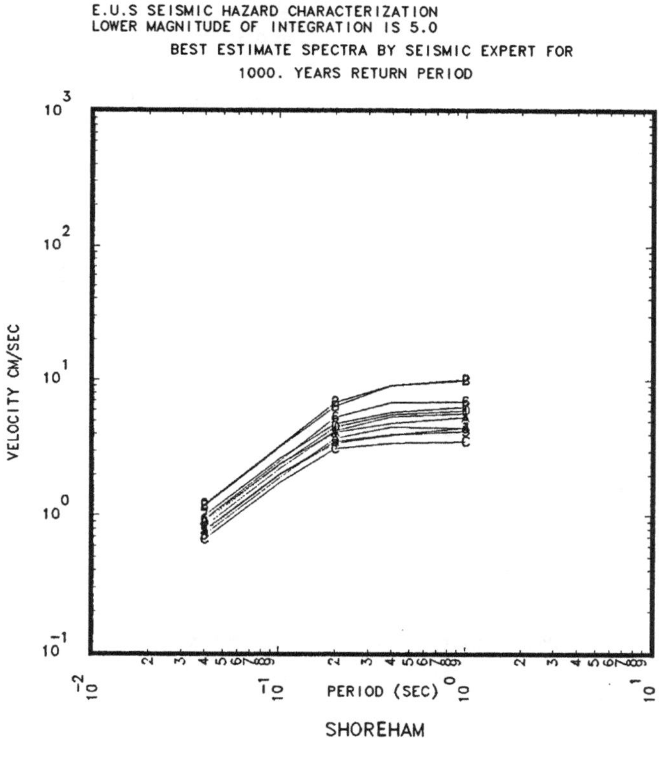

E.U.S SEISMIC HAZARD CHARACTERIZATION
LOWER MAGNITUDE OF INTEGRATION IS 5.0

BEST ESTIMATE SPECTRA BY SEISMIC EXPERT FOR
1000. YEARS RETURN PERIOD

VELOCITY CM/SEC

10^3 10^2 10^1 10^0 10^{-1}

10^{-2} 2 3 4 5 6 7 8 9 10^{-1} 2 3 4 5 6 7 8 9 10^0 2 3 4 5 6 7 8 9 10^1

PERIOD (SEC)

SHOREHAM

Figure 2.15.6 The 1000 year return period BEUHS per S-Expert aggregated over all G-Experts for the Shoreham site. Plot symbols are given in Table 2.0.

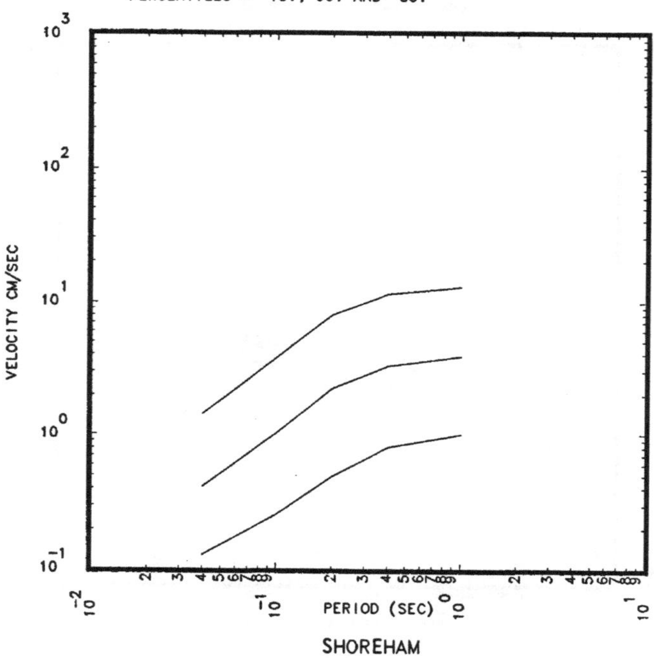

Figure 2.15.7 500 year return period CPUHS for the 15th, 50th and 85th percentiles aggregated over all S and G-Experts for the Shoreham site.

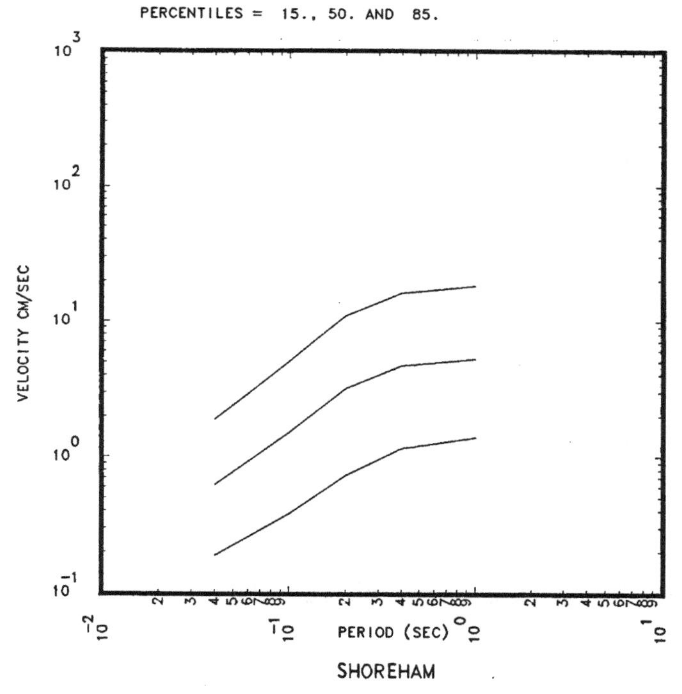

Figure 2.15.8 1000 year return period CPUHS for the 15th, 50th and 85th percentile aggregated over all S and G-Experts for the Shoreham site.

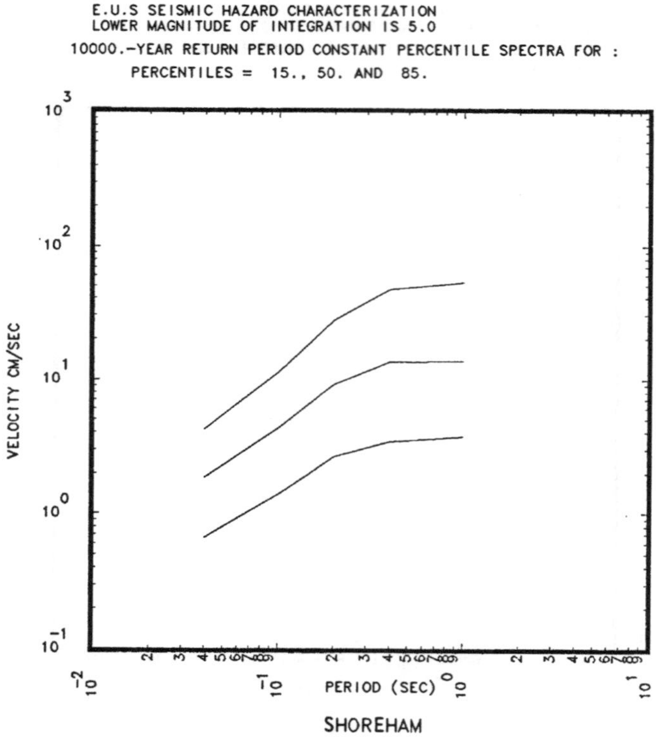

E.U.S SEISMIC HAZARD CHARACTERIZATION
LOWER MAGNITUDE OF INTEGRATION IS 5.0
10000.-YEAR RETURN PERIOD CONSTANT PERCENTILE SPECTRA FOR :
PERCENTILES = 15., 50. AND 85.

VELOCITY CM/SEC

PERIOD (SEC)

SHOREHAM

Figure 2.15.9 10000 year return period CPUHS for the 15th, 50th and 85th percentiles aggregated over all S and G-Experts for the Shoreham site.

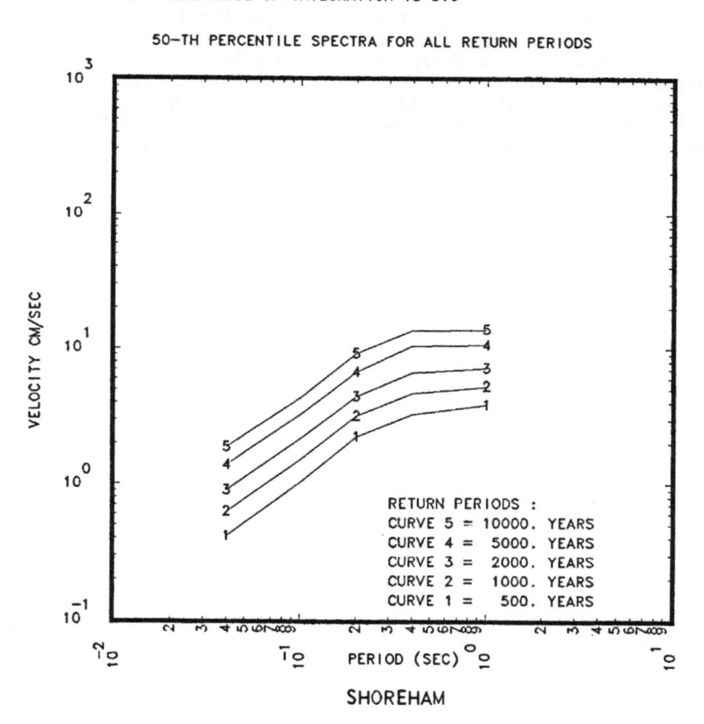

Figure 2.15.10 Comparison of the 50th percentile CPUHS for return periods of 500, 1000, 2000, 5000 and 10000 years for the Shoreham site.

2.16 SUSQUEHANNA

The Susquehanna site is a rock site and represented by the symbol "G" in Fig. 1.1. As noted in Section 2.0 the Susquehanna site was located in Region 1, but could have also been put into Region 2 or 3. The sensitivity to this placement is discussed in Section 3. Table 2.16.1 and Figs. 2.16.1 to 2.16.10 give the basic results for the Susquehanna site. The AMHC is about the same as the 85th percentile CPHC. We see from Fig. 2.16.2 that there is a relatively wide diversity of opinion between the S-Experts.

We see from Table 2.16.1 that for most S-Experts the zone which contains the site also is the zone which contributes most to the BEHC. For these S-Experts, the spread between the G-Experts' BEHCs are similar to that shown in Fig. 2.1.11. However, for S-Experts 2,5 and 6 the zonation and seismicity parameters are such that, when combined with G-Expert 5 GM model, larger more distant earthquakes contribute significantly to the BEHCs, when combined with G-Expert 5 GM model, and the spread between G-Expert 5's BEHCs and the other G-Experts' BEHCs for S-Experts 2,5,and 6 becomes much larger as illustrated in Fig. 2.16.11.

TABLE 2.16.1

MOST IMPORTANT ZONES PER S-EXPERT
FOR SUSQUEHANNA

SITE SOIL CATEGORY ROCK

S-XPT NUM.	MOST ZONE		ZONES CONTRIBUTING MOST SIGNIFICANTLY TO THE PGA BEHC AND % OF CONTRIBUTION AT LOW PGA(0.125G)				AT HIGH PGA(0.60G)			
1	ZONE 4	ZONE ID:	ZONE 4	ZONE 20	ZONE 21	ZONE 19	ZONE 4	ZONE 20	ZONE 21	ZONE 1
		% CON T:	76.	14.	4.	2.	95.	3.	1.	0.
2	COMP. ZO	ZONE ID:	ZONE 28	ZONE 32	ZONE 31	COMP. ZON	ZONE 28	COMP. ZON	ZONE 32	ZONE 31
		% CON T:	59.	19.	10.	8.	57.	24.	16.	3.
3	ZONE 5	ZONE ID:	ZONE 5	ZONE 4	ZONE 2	ZONE 11	ZONE 5	ZONE 4	COMP. ZON	ZONE 2
		% CONT.:	85.	4.	4.	4.	99.	0.	0.	0.
4	ZONE 12	ZONE ID:	ZONE 12	ZONE 16	ZONE 11	ZONE 19	ZONE 12	ZONE 16	ZONE 11	ZONE 8
		% CONT.:	64.	17.	7.	3.	97.	1.	1.	0.
5	COMP. ZO	ZONE ID:	ZONE 6	ZONE 1	ZONE 3	ZONE 5	ZONE 1	ZONE 6	COMP. ZON	ZONE 3
		% CONT.:	86.	9.	1.	1.	72.	19.	6.	2.
6	COMP. ZO	ZONE ID:	ZONE 6	ZONE 7	ZONE 3	COMP. ZON	ZONE 6	COMP. ZON	ZONE 7	ZONE 3
		% CON T:	54.	22.	9.	5.	76.	11.	11.	1.
7	ZONE 7	ZONE ID:	ZONE 7	ZONE 7	ZONE 41	ZONE 13	ZONE 7	ZONE 41	ZONE 17	ZONE 13
		% CON T:	67.	9.	6.	6.	97.	2.	0.	0.
10	ZONE 5	ZONE ID:	ZONE 5	ZONE 19 =	ZONE 4B	ZONE 6	ZONE 5	ZONE 19 =	ZONE 4B	ZONE 6
		% CON T:	70.	12.	5.	4.	93.	6.	1.	0.
11	ZONE 5	ZONE ID:	ZONE 5	ZONE 3	ZONE 4	CZ = ZONE	ZONE 5	CZ = ZONE	ZONE 4	ZONE 3
		% CONT.:	73.	11.	7.	6.	96.	2.	2.	1.
12	ZONE 27	ZONE ID:	ZONE 27	ZONE 32	ZONE 31	ZONE 34	ZONE 27	ZONE 32	ZONE 31	ZONE 34
		% CONT.:	41.	27.	22.	7.	90.	10.	0.	0.
13	CZ 15	ZONE ID:	CZ 15	ZONE 10	ZONE 12	ZONE 11	CZ 15	ZONE 10	ZONE 12	ZONE 11
		% CONT.:	58.	17.	12.	8.	98.	2.	0.	0.

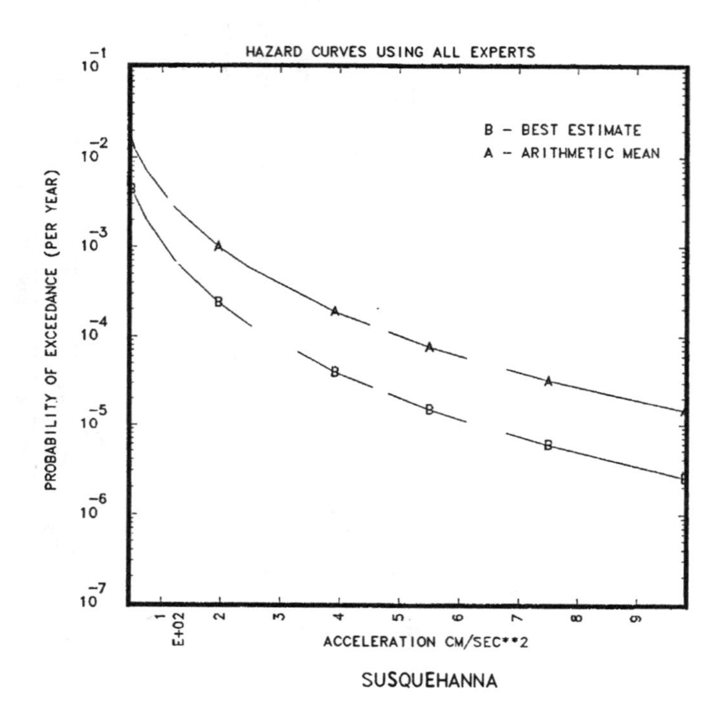

Figure 2.16.1 Comparison of the BEHC and AMHC aggregated over all S and G-Experts for the Susquehanna site.

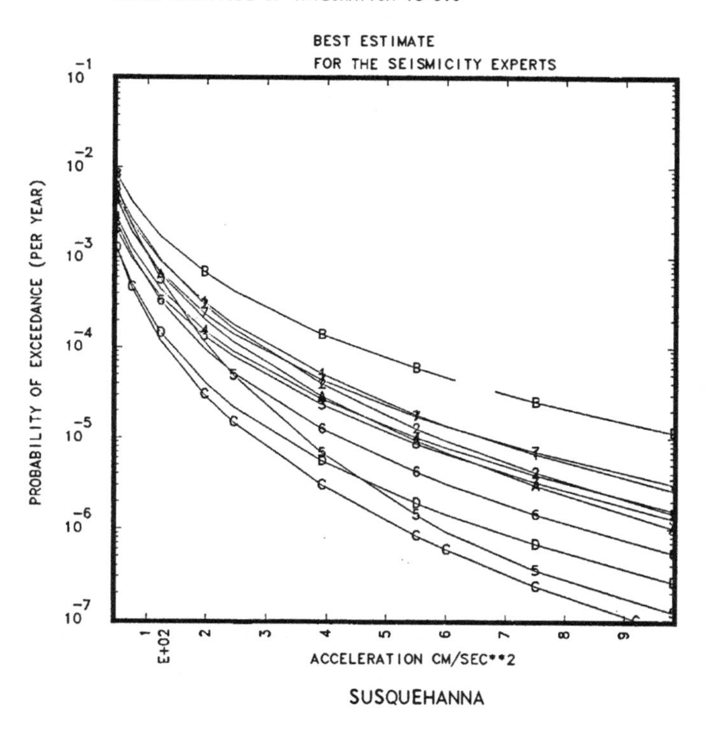

Figure 2.16.2 BEHCs per S-Expert combined over all G-Experts for the Susquehanna site. Plot symbols given in Table 2.0.

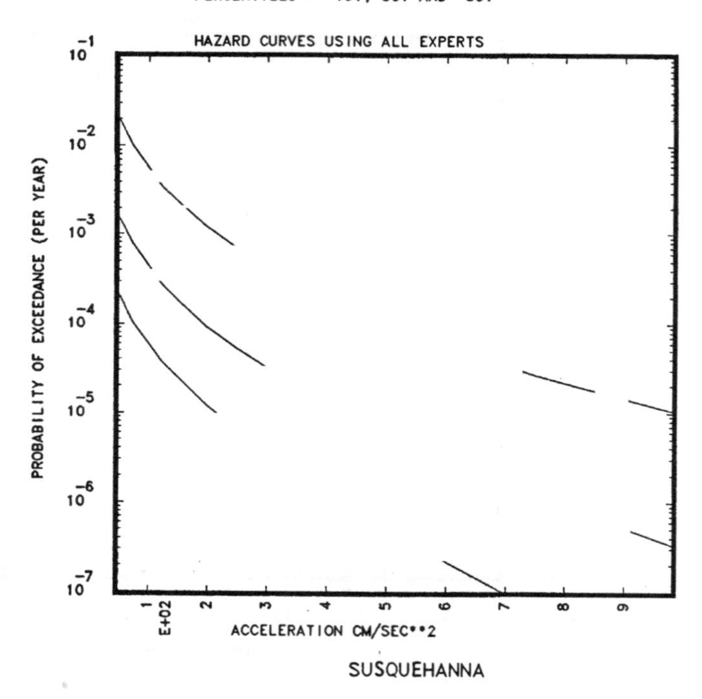

Figure 2.16.3 CPHCs for the 15th, 50th and 85th percentiles based on all S and G-Experts' input for the Susquehanna site.

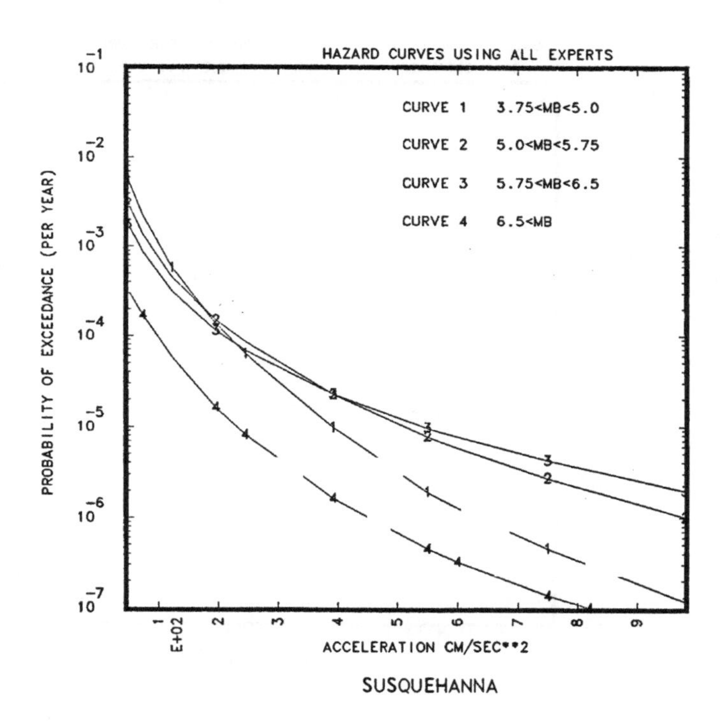

Figure 2.16.4 BEHCs which include only the contribution to the PGA hazard from earthquakes within the indicated magnitude range for the Susquehanna site.

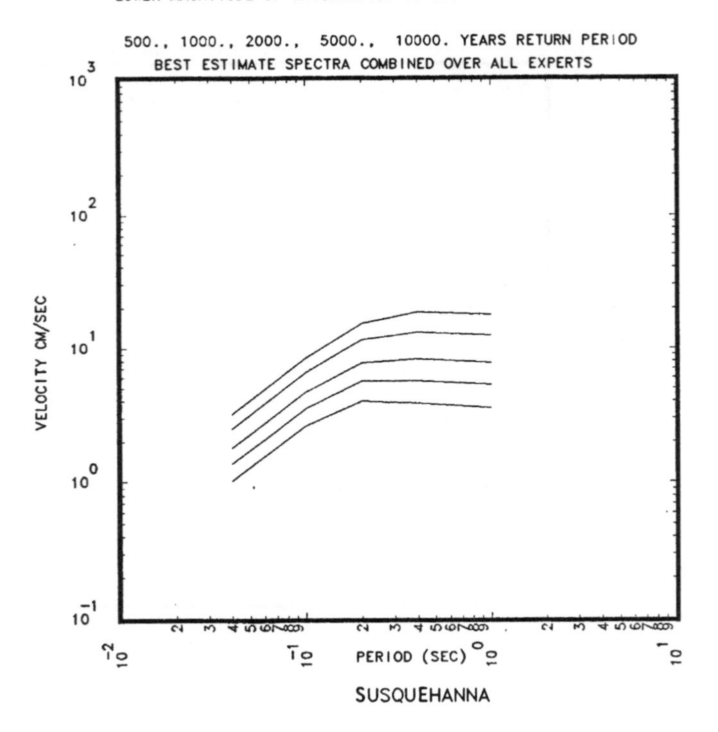

Figure 2.16.5 BEUHS for return periods of 500, 1000, 2000, 5000 and 10000 years aggregated over all S and G-Experts for the Susquehanna site.

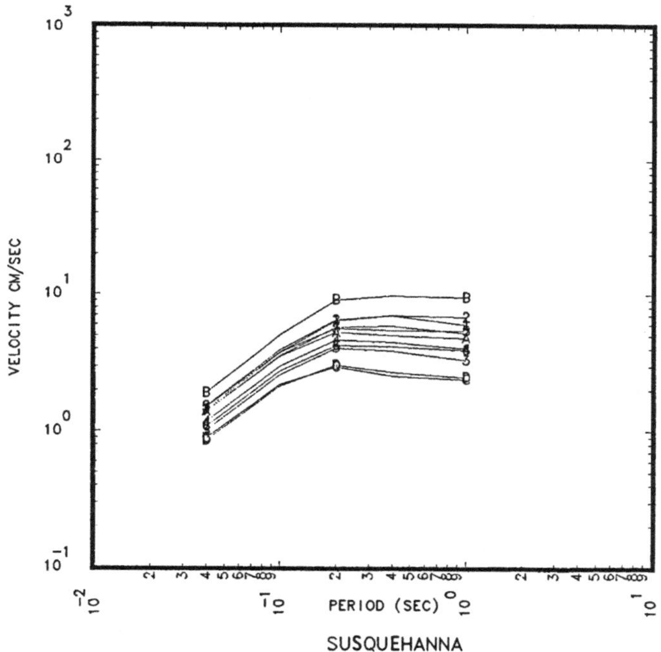

Figure 2.16.6 The 1000 year return period BEUHS per S-Expert aggregated over all G-Experts for the Susquehanna site. Plot symbols are given in Table 2.0.

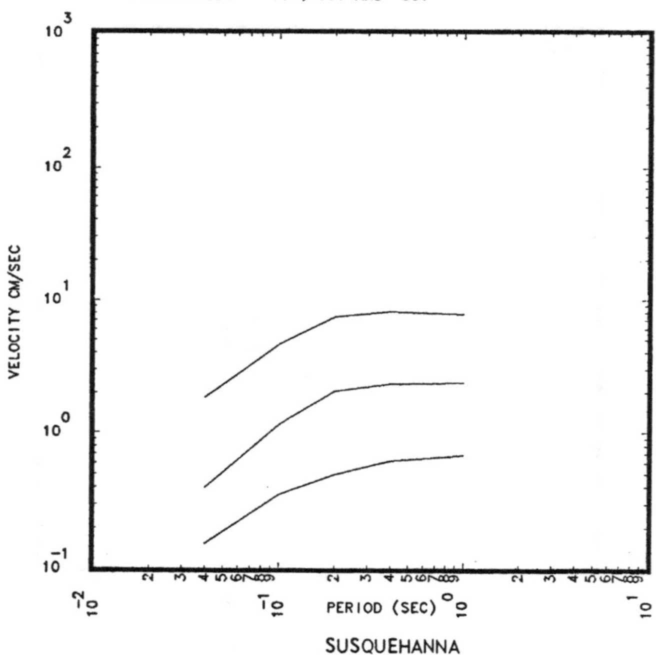

Figure 2.16.7 500 year return period CPUHS for the 15th, 50th and 85th percentiles aggregated over all S and G-Experts for the Susquehanna site.

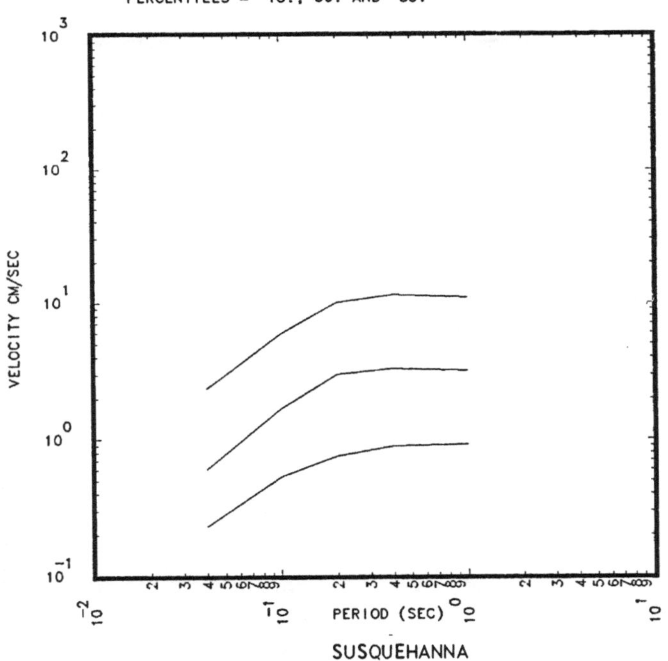

Figure 2.16.8 1000 year return period CPUHS for the 15th, 50th and 85th percentile aggregated over all S and G-Experts for the Susquehanna site.

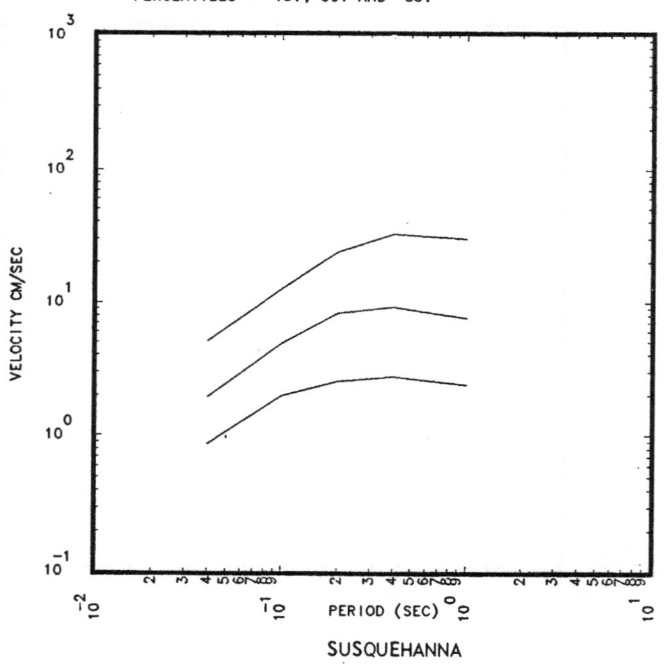

Figure 2.16.9 10000 year return period CPUHS for the 15th, 50th and 85th percentiles aggregated over all S and G-Experts for the Susquehanna site.

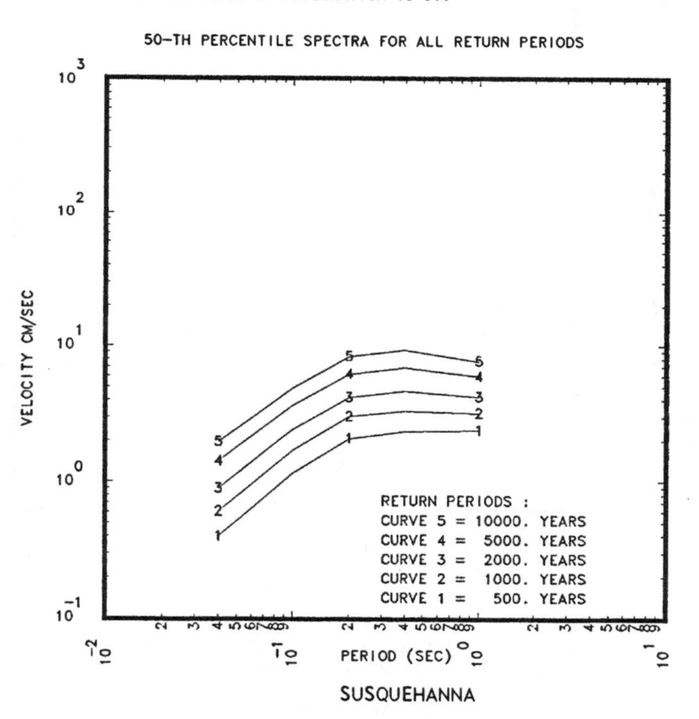

Figure 2.16.10 Comparison of the 50th percentile CPUHS for return periods of 500, 1000, 2000, 5000 and 10000 years for the Susquehanna site.

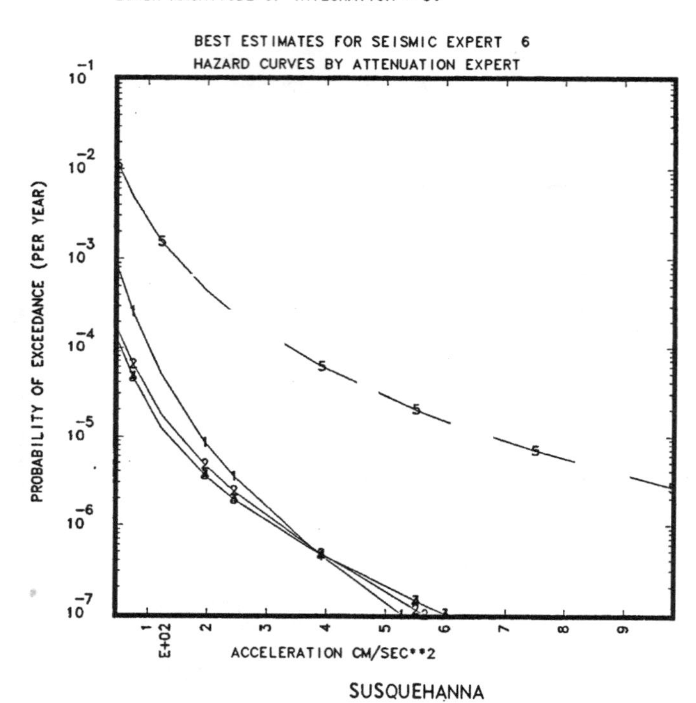

Figure 2.16.11. Comparison of the BEHCs for PGA per G-Expert for S-Expert 6's input for the Susquehanna site. The spread between G-Expert 5's BEHC and the other G-Experts' BEHC is somewhat larger than typical for rock sites in region 1. The spread between the Experts' BEHCs shown is typical for S-Experts 2, 5, and 6 at the Susquehanna site.

2.17 THREE MILE ISLAND

The Three Mile Island site is a rock site and represented by the symbol "H" in Fig. 1.1. As discussed in Section 2.0, the Three Mile Island site could be considered in either region 1,2 or 3. The results given in this section are based on using the region 1 GM models. In Section 3 we discuss the impact that this assumption has on the results.

Table 2.17.1 and Figs. 2.17.1 to 2.17.10 give the basic results for the Three Mile Island site. The AMHC is about as the 85th percentile CPHC.

We see from Table 2.17.1 that the zone which contains the site is also the zone which contributes the most to BEHC for all S-Experts except S-Expert 12 where a nearby zone is most significant. Thus, the spread between the G-Experts' BEHCs per S-Expert is typical for rock sites and is similar to the spread shown in Fig. 2.1.11.

TABLE 2.17.1

MOST IMPORTANT ZONES PER S-EXPERT
FOR THREE MI. ISLAND

SITE SOIL CATEGORY ROCK

S-XPT NO.	HOST ZONE		ZONES CONTRIBUTING MOST SIGNIFICANTLY TO THE PGA BEHC AND % OF CONTRIBUTION AT LOW PGA(0.125G)				AT HIGH PGA(0.60G)			
1	ZONE 4	ZONE ID:	ZONE 4	ZONE 20	ZONE 1	ZONE 19	ZONE 4	ZONE 1	ZONE 20	ZONE 21
		% CONT.:	83.	7.	3.	2.	98.	1.	1.	0.
2	ZONE 28	ZONE ID:	ZONE 28	ZONE 30	ZONE 32	ZONE 27	ZONE 28	ZONE 32	COMP. ZON	ZONE 30
		% CONT.:	81.	7.	5.	4.	96.	1.	1.	1.
3	ZONE 5	ZONE ID:	ZONE 5	ZONE 8A	ZONE [illegible]	ZONE 4	ZONE 5	ZONE 8A	COMP. ZON	ZONE 2
		% CONT.:	91.	2.	2.	2.	100.	0.	0.	0.
4	ZONE 11	ZONE ID:	ZONE 12	ZONE 11	ZONE 16	ZONE 8	ZONE 11	ZONE 12	ZONE 8	ZONE 16
		% CONT.:	37.	31.	11.	10.	54.	38.	7.	0.
5	ZONE 1	ZONE ID:	ZONE 6	ZONE 1	ZONE 9	ZONE 12	ZONE 1	ZONE 9	ZONE 6	COMP. Z
		% CONT.:	59.	29.	8.	1.	98.	1.	1.	0.
6	COMP. ZO	ZONE ID:	ZONE 6	ZONE 7	ZONE 15	ZONE 13	ZONE 6	COMP. ZON	ZONE 7	ZONE 15
		% CONT.:	72.	9.	6.	5.	93.	5.	1.	1.
7	ZONE 7	ZONE ID:	ZONE 7	ZONE 29	ZONE 17	ZONE 13	ZONE 7	ZONE 29	ZONE 10	ZONE 41
		% CONT.:	82.	6.	3.	3.	99.	0.	0.	0.
10	ZONE 4B	ZONE ID:	ZONE 5	ZONE 4B	ZONE 19 =	ZONE 6	ZONE 4B	ZONE 5	ZONE 19 =	ZONE
		% CONT.:	56.	31.	5.	2.	55.	43.	2.	0.
11	ZONE 5	ZONE ID:	ZONE 5	ZONE 3	CZ = ZONE	ZONE 4	ZONE 5	CZ = ZONE	ZONE 8	ZONE 3
		% CONT.:	86.	5.	3.	2.	99.	0.	0.	0.
12	ZONE 27	ZONE ID:	ZONE 32	ZONE 27	ZONE 31	ZONE 23A	ZONE 32	ZONE 27	ZONE 23A	ZONE 34
		% CONT.:	71.	18.	7.	1.	88.	12.	0.	0.
13	CZ 15	ZONE ID:	CZ 15	CZ 17	ZONE 10	ZONE 11	CZ 15	CZ 17	ZONE 10	ZONE 12
		% CONT.:	71.	11.	9.	3.	99.	0.	0.	0.

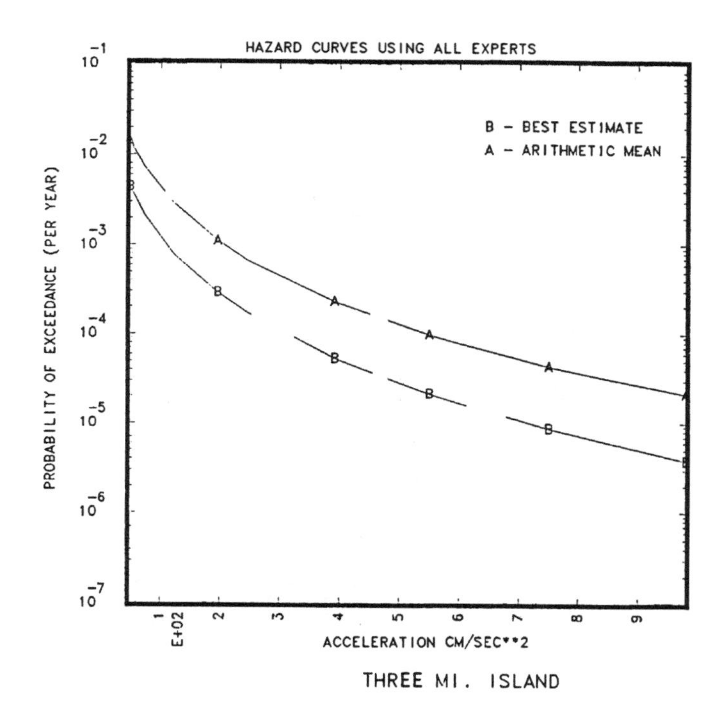

Figure 2.17.1 Comparison of the BEHC and AMHC aggregated over all S and G-Experts for the Three Mile Island site.

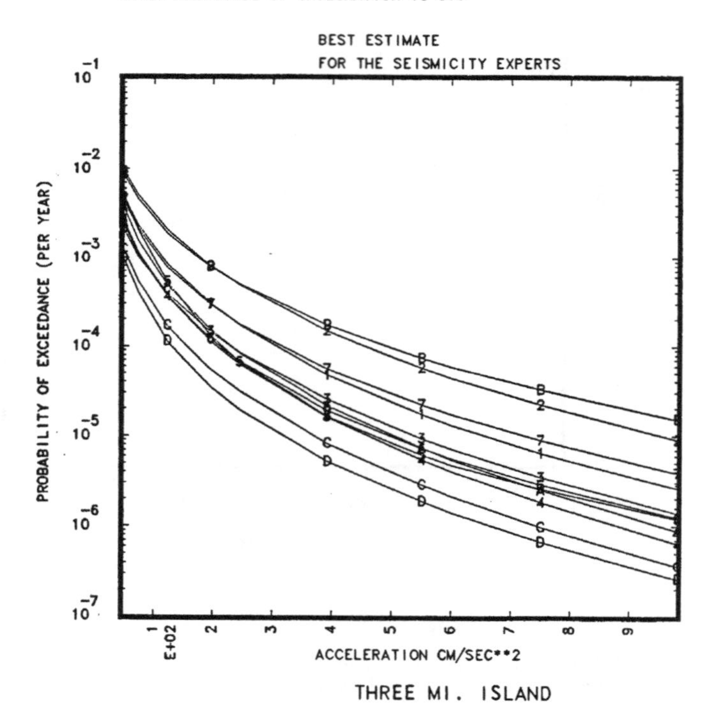

Figure 2.17.2 BEHCs per S-Expert combined over all G-Experts for the Three Mile Island site. Plot symbols given in Table 2.0.

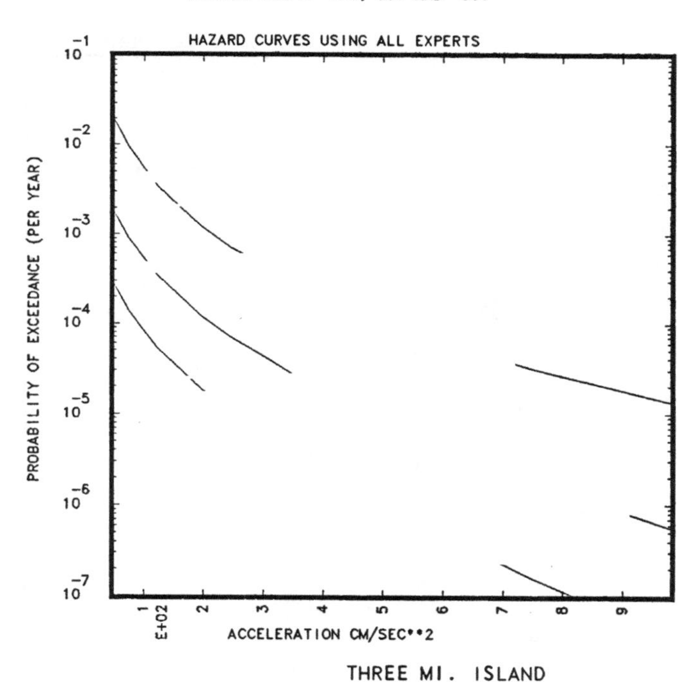

Figure 2.17.3 CPHCs for the 15th, 50th and 85th percentiles based on all S and G-Experts' input for the Three Mile Island site.

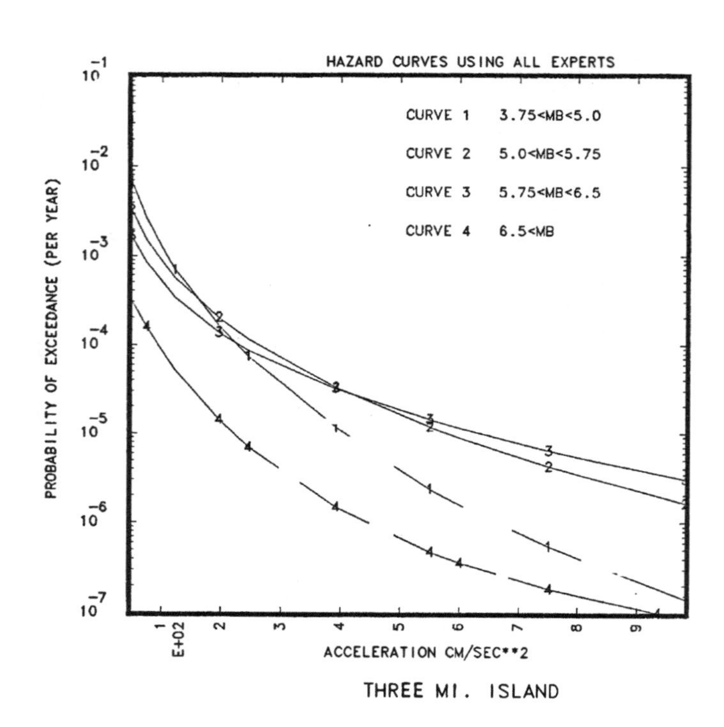

Figure 2.17.4 BEHCs which include only the contribution to the PGA hazard from earthquakes within the indicated magnitude range for the Three Mile Island site.

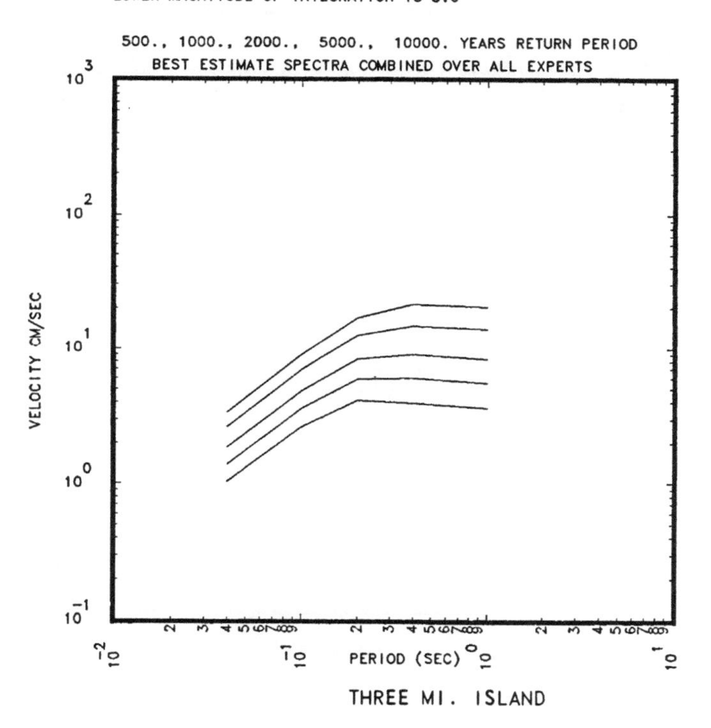

Figure 2.17.5 BEUHS for return periods of 500, 1000, 2000, 5000 and 10000 years aggregated over all S and G-Experts for the Three Mile Island site.

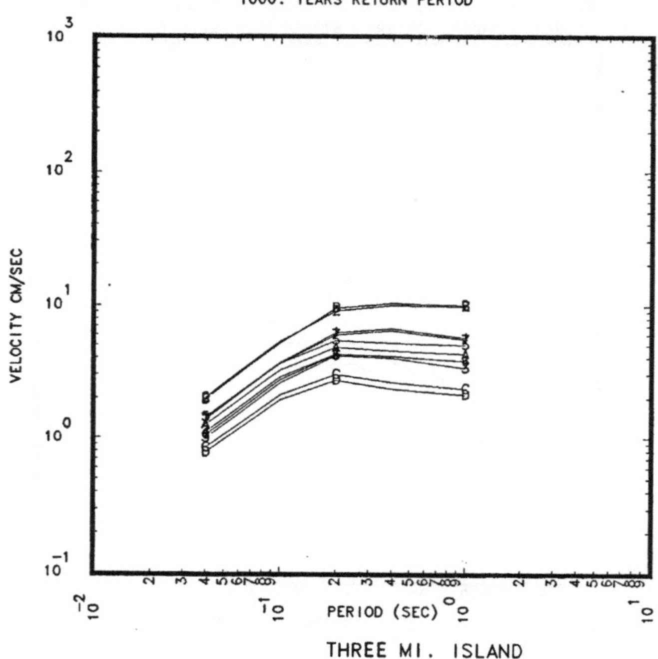

Figure 2.17.6 The 1000 year return period BEUHS per S-Expert aggregated over all G-Experts for the Three Mile Island site. Plot symbols are given in Table 2.0.

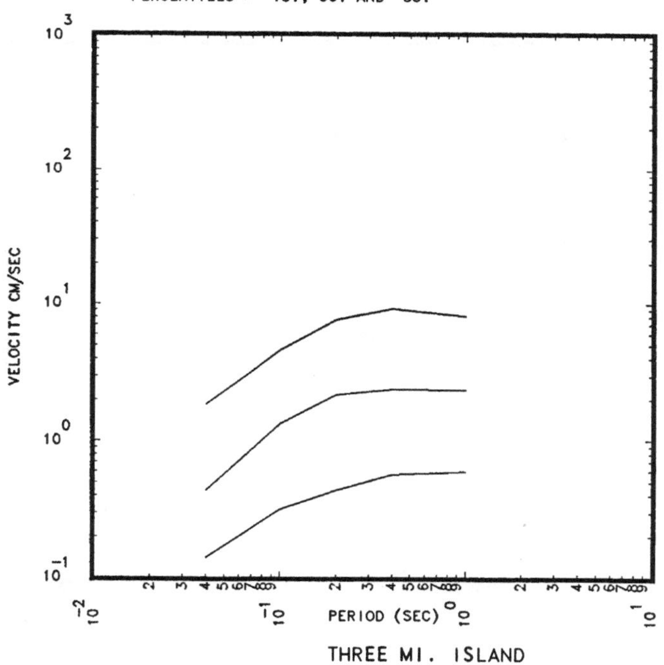

Figure 2.17.7 500 year return period CPUHS for the 15th, 50th and 85th percentiles aggregated over all S and G-Experts for the Three Mile Island site.

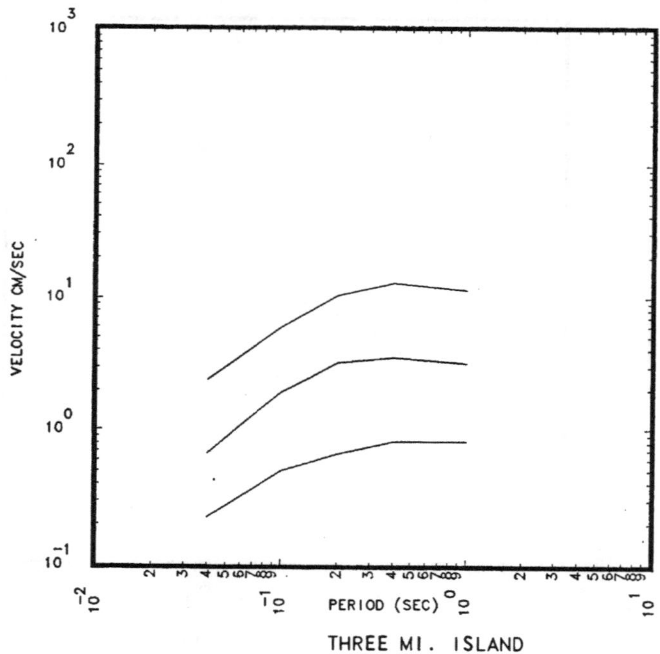

Figure 2.17.8 1000 year return period CPUHS for the 15th, 50th and 85th percentile aggregated over all S and G-Experts for the Three Mile Island site.

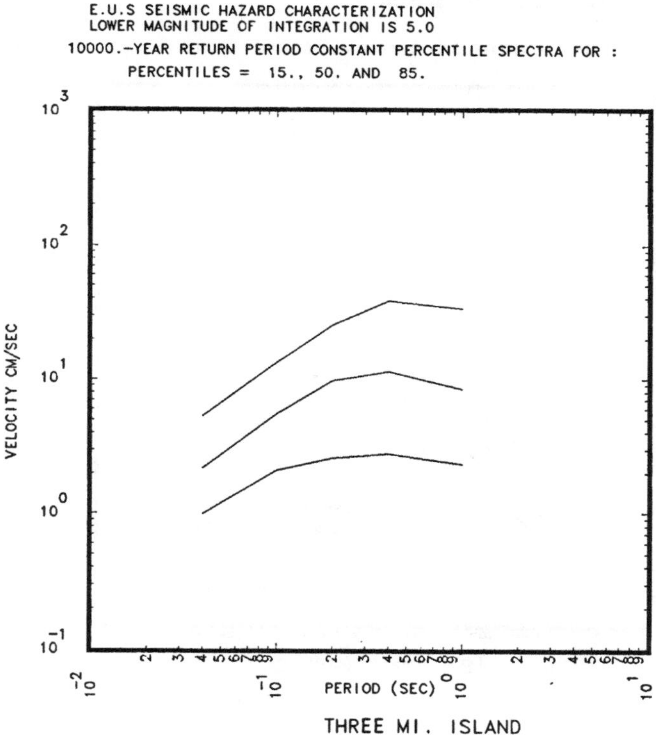

E.U.S SEISMIC HAZARD CHARACTERIZATION
LOWER MAGNITUDE OF INTEGRATION IS 5.0
10000.-YEAR RETURN PERIOD CONSTANT PERCENTILE SPECTRA FOR :
PERCENTILES = 15., 50. AND 85.

VELOCITY CM/SEC
PERIOD (SEC)
THREE MI. ISLAND

Figure 2.17.9 10000 year return period CPUHS for the 15th, 50th and 85th percentiles aggregated over all S and G-Experts for the Three Mile Island site.

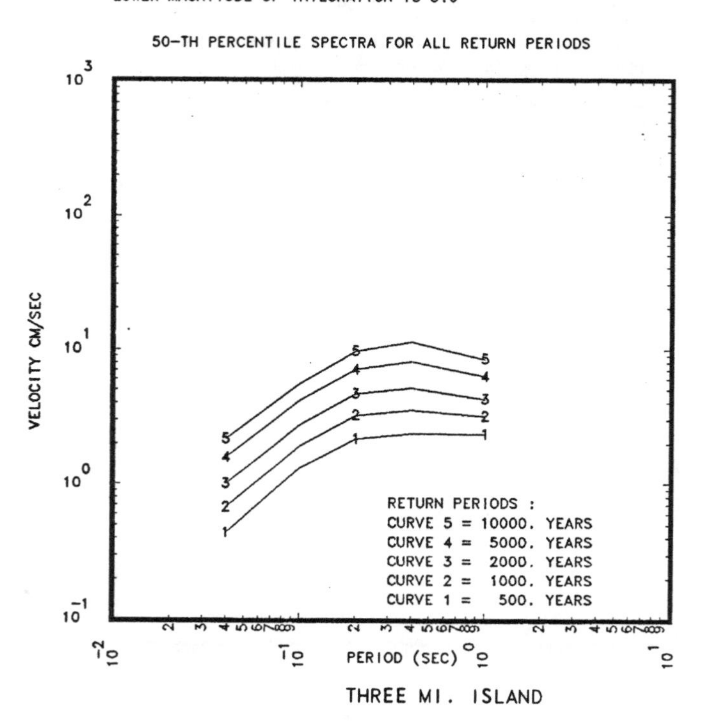

Figure 2.17.10 Comparison of the 50th percentile CPUHS for return periods of 500, 1000, 2000, 5000 and 10000 years for the Three Mile Island site.

2.18 VERMONT YANKEE

The Vermont Yankee site is a rock site and represented by the symbol "I" in Fig. 1.1. Table 2.18.1 and Figs. 2.18.1 to 2.18.10 give the basic results for the Vermont Yankee site. The AMHC is higher than the 85th percentile CPHC, particularly at high g values indicating the existence of some high outliers.

We see from Table 2.18.1 that for S-Experts 1,2,3,5,6 and 12 the zone which contains the site is also the zone which contributes most to the BEHC. For these S-Experts the spread between the G-Experts' BEHC is typical for rock sites in Region 1 and similar to that shown in fig. 2.1.11. For S-Experts 4,7,10,11 and 13 more distant zones make the most significant contribution to the BEHC. For these S-Experts the spread between G-Expert 5's BEHC and the other G-Experts BEHCs, per S-Expert, is larger as shown in Fig. 2.18.11. For S-Expert 4 the spread between G-Expert 5's BEHC and the other G-Experts BEHC is even larger, similar to the spread shown in Fig. 2.1.12. Given the importance of distant zones for a number of S-Experts it is not unexpected that, as can be seen from Fig. 2.18.4, the BEHC would not significantly change if earthquakes in the range 3.75 to 5 were included.

TABLE 2.18.1

MOST IMPORTANT ZONES PER S-EXPERT
FOR VERMONT YANKEE

SITE SOIL CATEGORY ROCK

S-XPT NUM.	MOST ZONE		ZONES CONTRIBUTING MOST SIGNIFICANTLY TO THE PGA BEHC AND % OF CONTRIBUTION AT LOW PGA(0.125G)				AT HIGH PGA(0.60G)			
1	ZONE 22	ZONE ID:	ZONE 22	ZONE 20	ZONE 21	ZONE 4	ZONE 22	ZONE 20	ZONE 21	ZONE 4
		% CONT.:	45.	38.	14.	1.	67.	27.	6.	0.
2	ZONE 31	ZONE ID:	ZONE 31	ZONE 32	ZONE 28	COMP. ZON	ZONE 31	ZONE 32	COMP. ZON	ZONE 28
		% CON T:	62.	35.	2.	1.	68.	31.	0.	0.
3	ZONE 4	ZONE ID:	ZONE 4	ZONE 2	ZONE 3	ZONE 5	ZONE 4	ZONE 2	ZONE 3	ZONE 5
		% CONT.:	77.	12.	9.	2.	95.	4.	1.	0.
4	COMP. ZO	ZONE ID:	ZONE 18	ZONE 16	ZONE 20	ZONE 19	ZONE 18	ZONE 16	ZONE 20	ZONE 17
		% CONT.:	50.	21.	9.	8.	75.	17.	3.	2.
5	ZONE 1	ZONE ID:	ZONE 1	ZONE 6	ZONE 3	ZONE 4	ZONE 1	ZONE 3	ZONE 4	COMP. ZON
		% CONT.:	53.	36.	7.	4.	95.	4.	1.	0.
6	ZONE 5	ZONE ID:	ZONE 5	ZONE 7	ZONE 6	ZONE 3	ZONE 5	ZONE 7	ZONE 6	ZONE 3
		% CONT.:	41.	24.	20.	14.	65.	16.	16.	2.
7	ZONE 24	ZONE ID:	ZONE 19	ZONE 17	ZONE 24	ZONE 18	ZONE 19	ZONE 17	ZONE 24	ZONE 18
		% CONT.:	45.	24.	11.	10.	66.	16.	15.	1.
10	ZONE 3	ZONE ID:	ZONE 23	ZONE 6	ZONE 21	ZONE 22	ZONE 23	ZONE 22	ZONE 3	ZONE 21
		% CONT.:	37.	16.	13.	10.	69.	9.	9.	6.
11	CZ = ZON	ZONE ID:	ZONE 3	ZONE 5	ZONE 2	CZ = ZONE	ZONE 3	ZONE 5	CZ = ZONE	ZONE 4
		% CONT.:	49.	16.	11.	9.	37.	33.	14.	9.
12	ZONE 32	ZONE ID:	ZONE 32	ZONE 31	ZONE 33	ZONE 34	ZONE 32	ZONE 33	ZONE 31	ZONE 34
		% CONT.:	43.	23.	20.	11.	82.	12.	3.	2.
13	CZ 15	ZONE ID:	ZONE 10	ZONE 12	ZONE 11	CZ 15	ZONE 10	CZ 15	ZONE 12	ZONE 11
		% CONT.:	61.	16.	12.	11.	73.	21.	3.	3.

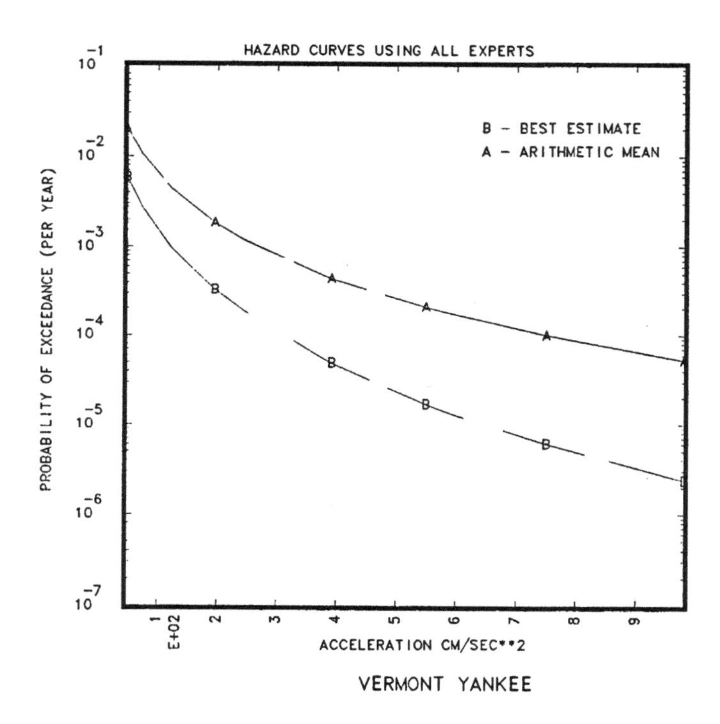

Figure 2.18.1 Comparison of the BEHC and AMHC aggregated over all S and G-Experts for the Vermont Yankee site.

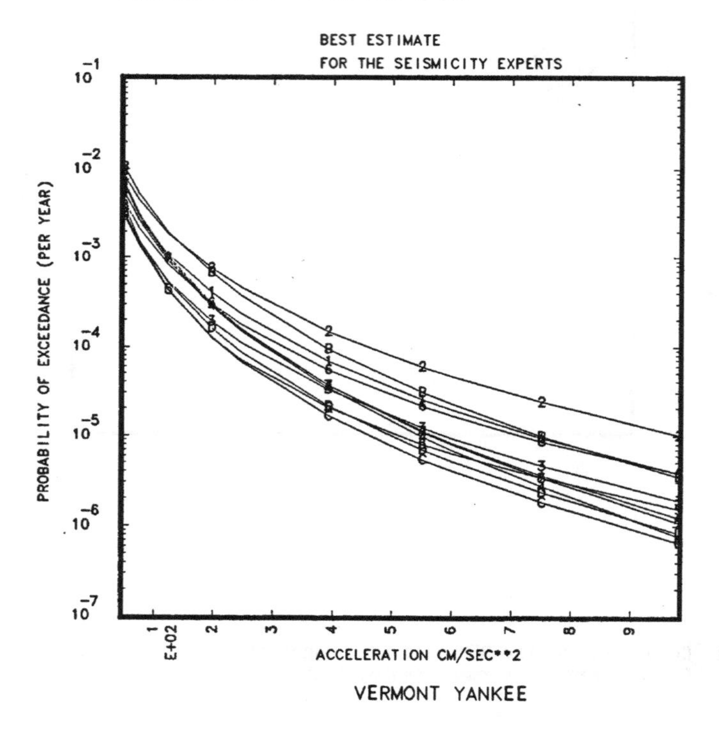

Figure 2.18.2 BEHCs per S-Expert combined over all G-Experts for the Vermont Yankee site. Plot symbols given in Table 2.0.

E.U.S SEISMIC HAZARD CHARACTERIZATION
LOWER MAGNITUDE OF INTEGRATION IS 5.0
PERCENTILES = 15., 50. AND 85.

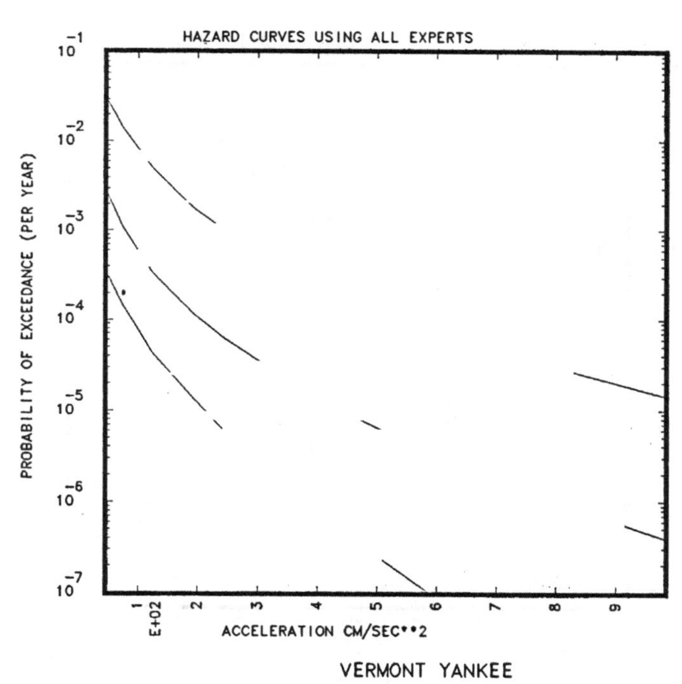

Figure 2.18.3 CPHCs for the 15th, 50th and 85th percentiles based on all S and G-Experts' input for the Vermont Yankee site.

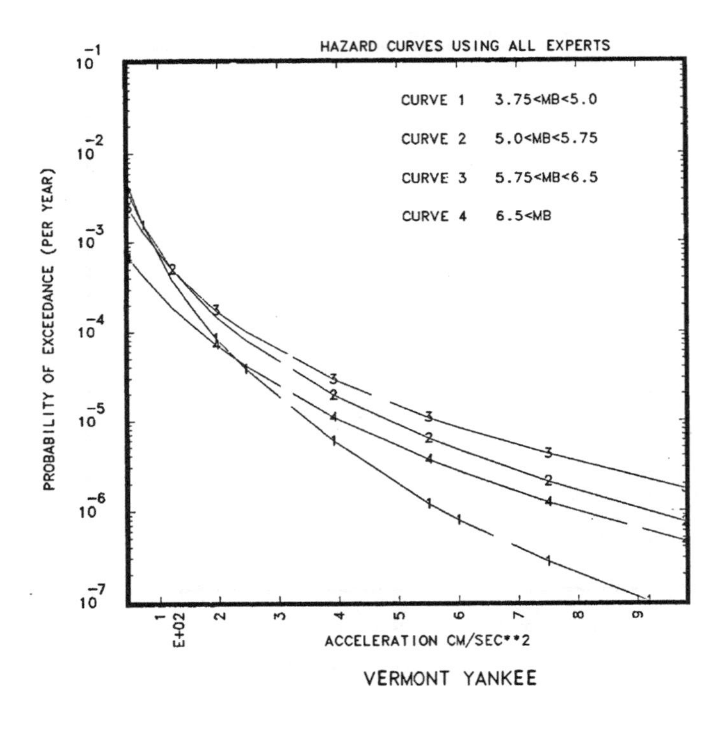

Figure 2.18.4 BEHCs which include only the contribution to the PGA hazard from earthquakes within the indicated magnitude range for the Vermont Yankee site.

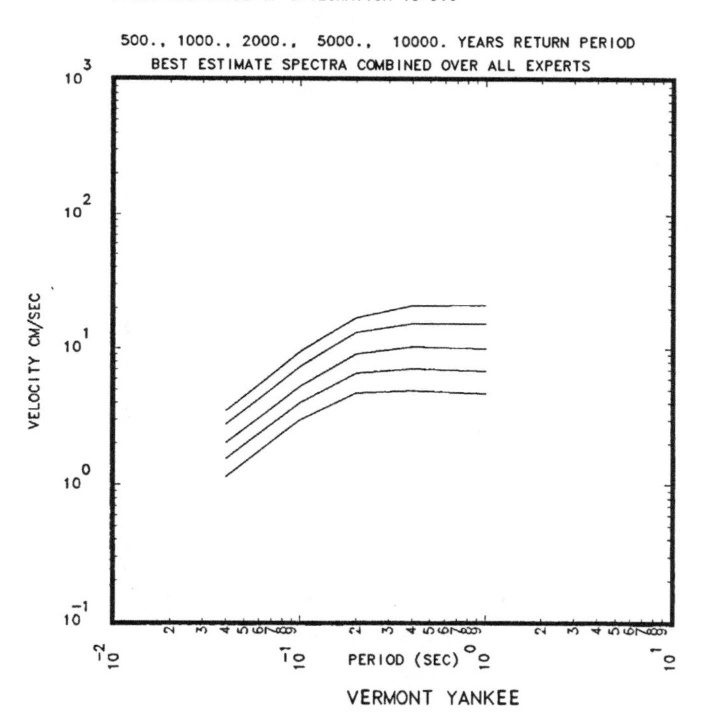

Figure 2.18.5 BEUHS for return periods of 500, 1000, 2000, 5000 and 10000 years aggregated over all S and G-Experts for the Vermont Yankee site.

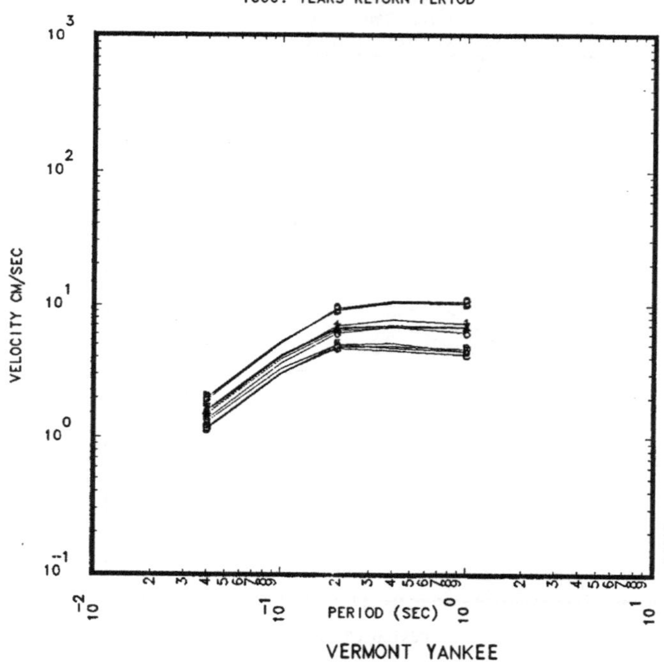

Figure 2.18.6 The 1000 year return period BEUHS per S-Expert aggregated over all G-Experts for the Vermont Yankee site. Plot symbols are given in Table 2.0.

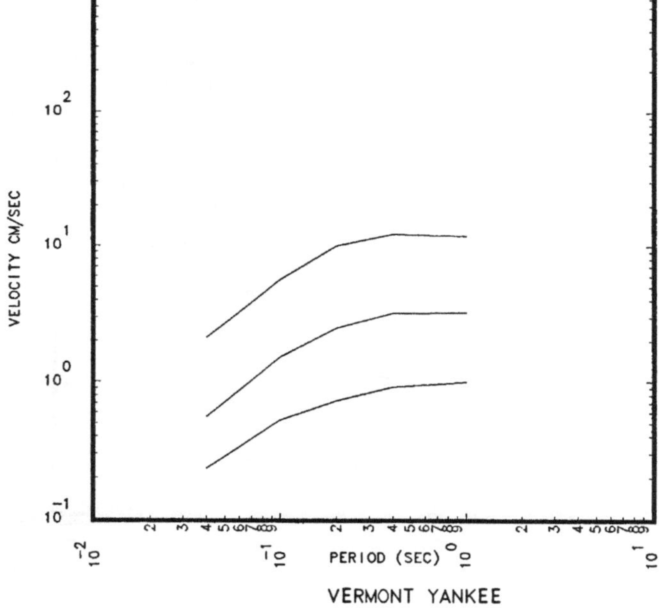

Figure 2.18.7 500 year return period CPUHS for the 15th, 50th and 85th percentiles aggregated over all S and G-Experts for the Vermont Yankee site.

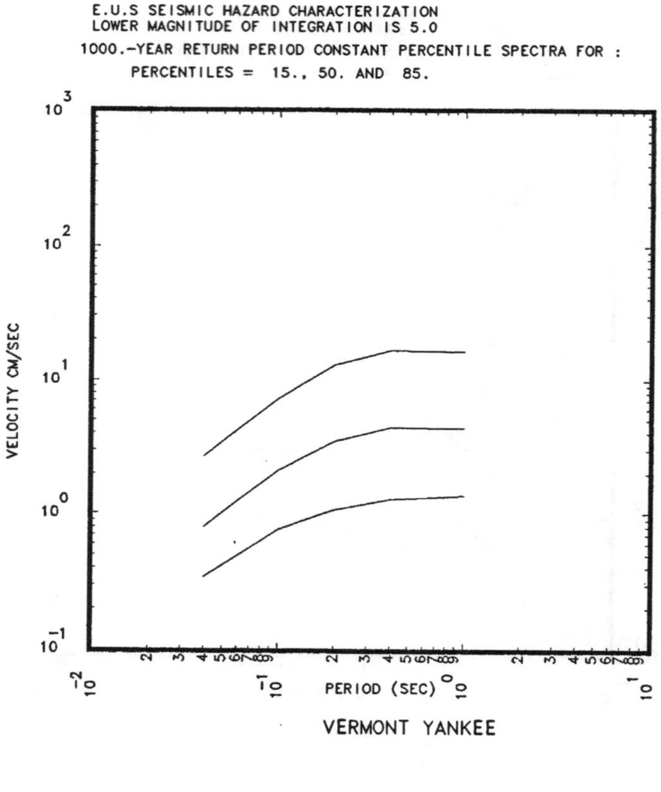

E.U.S SEISMIC HAZARD CHARACTERIZATION
LOWER MAGNITUDE OF INTEGRATION IS 5.0
1000.-YEAR RETURN PERIOD CONSTANT PERCENTILE SPECTRA FOR :
PERCENTILES = 15., 50. AND 85.

VELOCITY CM/SEC

PERIOD (SEC)

VERMONT YANKEE

Figure 2.18.8 1000 year return period CPUHS for the 15th, 50th and 85th percentile aggregated over all S and G-Experts for the Vermont Yankee site.

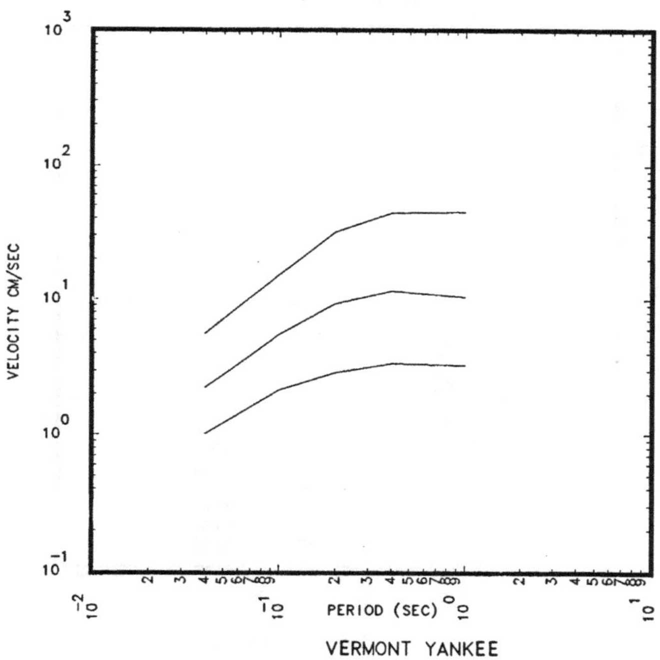

Figure 2.18.9 10000 year return period CPUHS for the 15th, 50th and 85th percentiles aggregated over all S and G-Experts for the Vermont Yankee site.

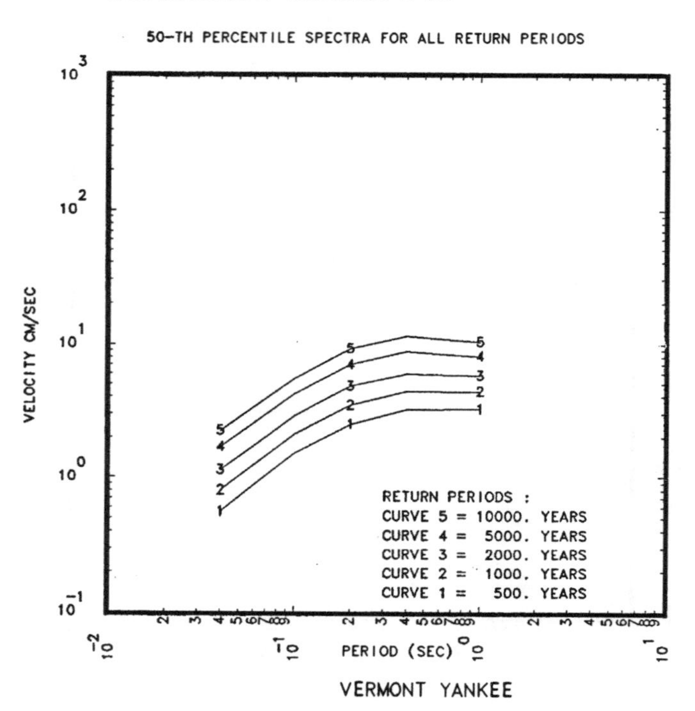

Figure 2.18.10 Comparison of the 50th percentile CPUHS for return periods of 500, 1000, 2000, 5000 and 10000 years for the Vermont Yankee site.

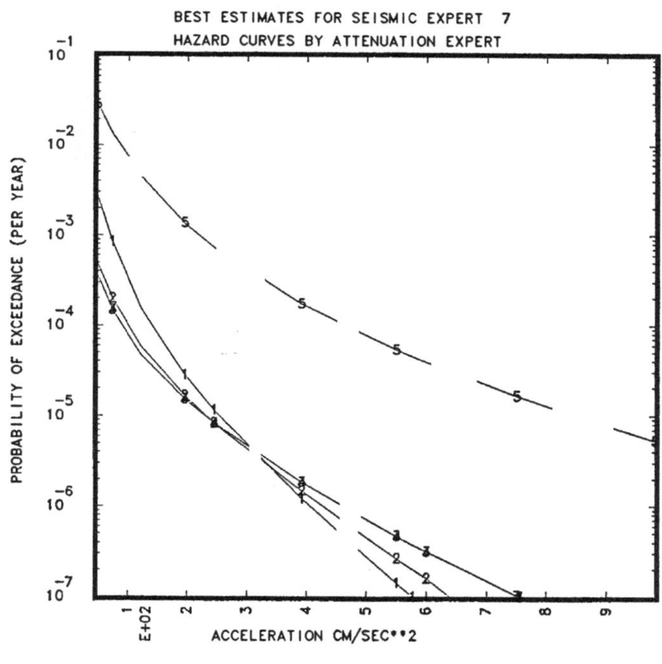

Figure 2.18.11. Comparison of BEHCs for PGA per G-Expert for S-Expert 7's input for the Vermont Yankee site. The spread between G-Expert 5's BEHC and the other G-Experts' BEHCs is somewhat larger than typical for rock sites in region 1. The spread shown is typical for S-Experts 7, 10, 11, and 13's input for the Vermont Yankee site.

2.19 YANKEE ROWE

The Yankee Rowe site's soil category is Till-2 and is represented by the symbol "J" in Fig. 1.1. Table 2.19.1 and Figs. 2.19.1 to 2.19.10 give the basic results for the Yankee Rowe site. The BEHC is about the same as the 50th percentile CPHC and the AMHC is about the same as the 85th percentile CPHC.

For all S-Experts, but S-Experts 4 and 11, the zone which contains the site is also the zone which contributes most to the BEHC. However, only for S-Experts 3,5 and 12 is the spread between the G-Experts' BEHCs, per S-Expert, similar to that at other soil sites. For S-Experts 1,2,6,7 and 13 the spread between G-Expert 5's BEHCs and the other G-Experts' BEHCs is somewhat larger, as shown in Fig. 2.19.11, indicating that larger earthquakes are somewhat more important than at other soil sites. Fig. 2.19.4 shows that small earthquakes would contribute significantly to the BEHC if they were to be included as expected by the fact that the zone which contains the site is also the zone which contributes most to the hazard.

The spread between G-Expert 5's BEHC as compared to the other G-Experts for S-Expert 4's input is even larger than illustrated in Fig. 2.19.11. However, it is not as large as shown in Fig. 2.1.12.

TABLE 2.19.1

MOST IMPORTANT ZONES PER S-EXPERT
FOR YANKEE ROWE

SITE SOIL CATEGORY TILL-2

S-XPT NUM.	MOST ZONE		ZONES CONTRIBUTING MOST SIGNIFICANTLY TO THE PGA BEHC AND % OF CONTRIBUTION AT LOW PGA(0.125G)				AT HIGH PGA(0.60G)			
1	ZONE 22	ZONE ID:	ZONE 22	ZONE 20	ZONE 21	ZONE 4	ZONE 22	ZONE 20	ZONE 21	ZONE 1
		% CONT.:	56.	34.	9.	1.	79.	19.	2.	0.
2	ZONE 31	ZONE ID:	ZONE 31	ZONE 32	ZONE 28	COMP. ZON	ZONE 31	ZONE 32	COMP. ZON	ZONE 2
		% CONT.:	75.	24.	1.	1.	85.	15.	0.	0.
3	ZONE 4	ZONE ID:	ZONE 4	ZONE 2	ZONE 3	ZONE 5	ZONE 4	ZONE 2	ZONE 3	ZONE 5
		% CONT.:	86.	9.	4.	1.	97.	3.	0.	0.
4	MP. ZO	ZONE ID:	ZONE 18	ZONE 17	ZONE 16	ZONE 20	ZONE 18	ZONE 16	ZONE 17	COMP.
		% CONT.:	34.	29.	22.	6.	43.	33.	17.	5.
5	ZONE 1	ZONE ID:	ZONE 1	ZONE 6	ZONE 3	ZONE 4	ZONE 1	ZONE 3	ZONE 4	COMP.
		% CONT.:	75.	17.	5.	3.	99.	1.	0.	0.
6	ZONE 6	ZONE ID:	ZONE 6	ZONE 7	ZONE 5	ZONE 3	ZONE 6	ZONE 5	ZONE 7	COMP.
		% CONT.:	35.	29.	27.	8.	50.	33.	16.	1.
7	ZONE 24	ZONE ID:	ZONE 17	ZONE 19	ZONE 24	ZONE 18	ZONE 24	ZONE 17	ZONE 19	ZONE 1
		% CONT.:	33.	28.	23.	9.	50.	30.	19.	1.
10	ZONE 3	ZONE ID:	ZONE 6	ZONE 23	ZONE 3	ZONE 21	ZONE 3	ZONE 23	ZONE 6	ZONE
		% CONT.:	28.	26.	15.	9.	51.	22.	18.	4.
11	CZ = ZON	ZONE ID:	ZONE 5	ZONE 3	ZONE 4	CZ = ZONE	ZONE 5	CZ = ZONE	ZONE 4	ZONE
		% CONT.:	39.	30.	12.	10.	73.	11.	8.	8.
12	ZONE 32	ZONE ID:	ZONE 32	ZONE 31	ZONE 33	ZONE 34	ZONE 32	ZONE 33	ZONE 31	ZONE 3
		% CONT.:	69.	14.	8.	6.	98.	1.	1.	0.
13	CZ 15	ZONE ID:	ZONE 10	CZ 15	ZONE 11	ZONE 12	CZ 15	ZONE 10	ZONE 11	ZONE 1
		% CONT.:	43.	31.	13.	13.	82.	14.	3.	1.

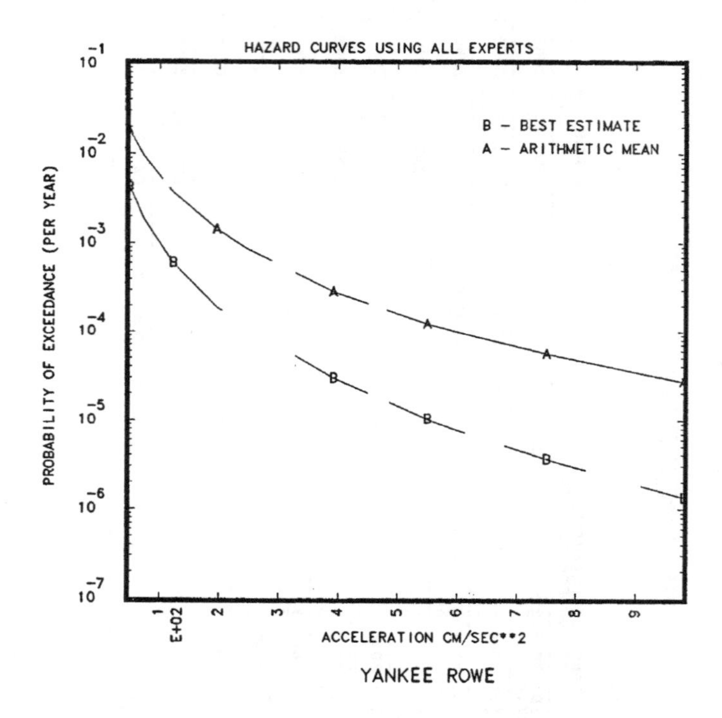

Figure 2.19.1 Comparison of the BEHC and AMHC aggregated over all S and G-Experts for the Yankee Rowe site.

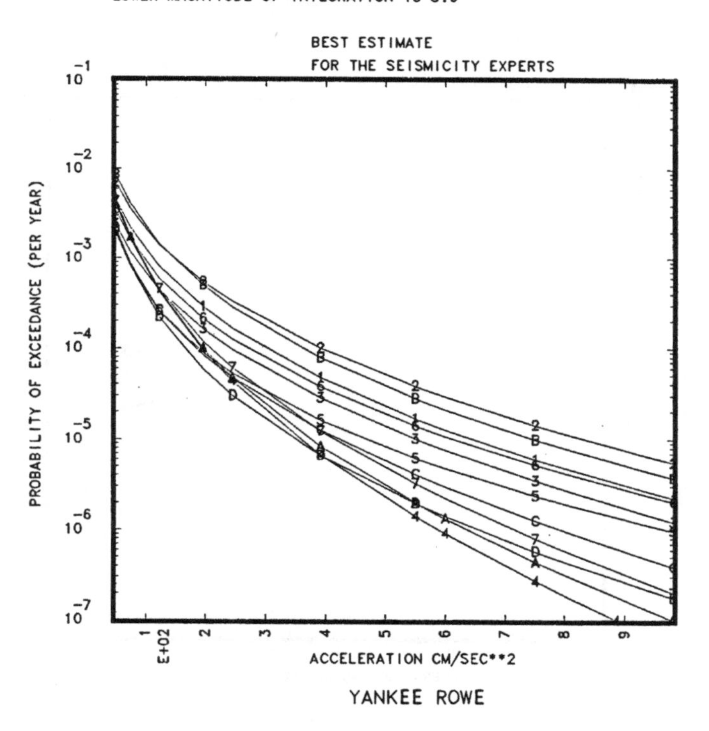

Figure 2.19.2 BEHCs per S-Expert combined over all G-ExPerts for the Yankee at Rowe site. Plot symbols given in Table 2.0.

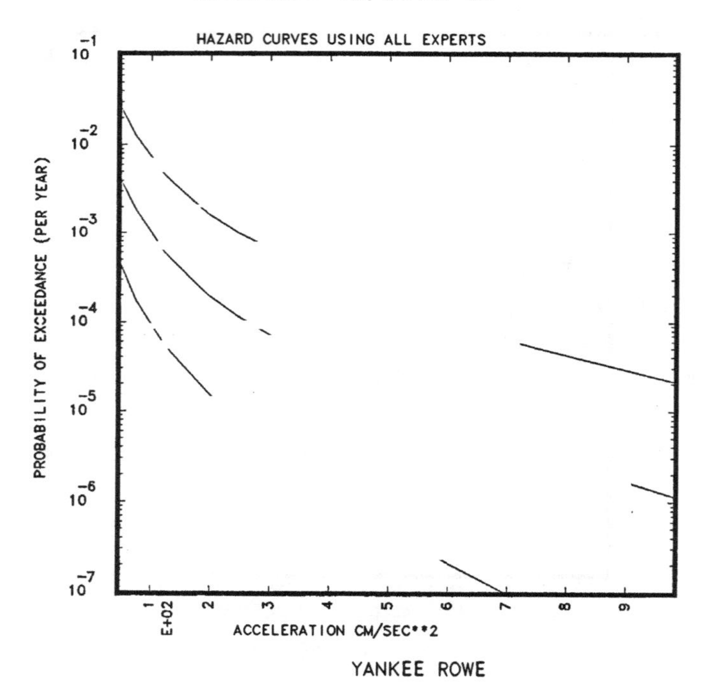

Figure 2.19.3 CPHCs for the 15th, 50th and 85th percentiles based on all S and G-Experts' input for the Yankee at Rowe site.

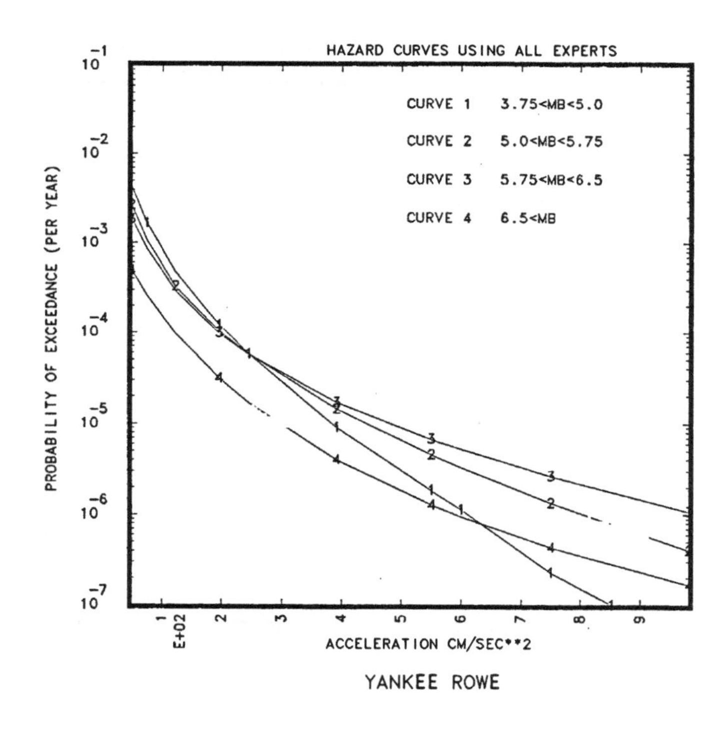

Figure 2.19.4 BEHCs which include only the contribution to the PGA hazard from earthquakes within the indicated magnitude range for the Yankee at Rowe site.

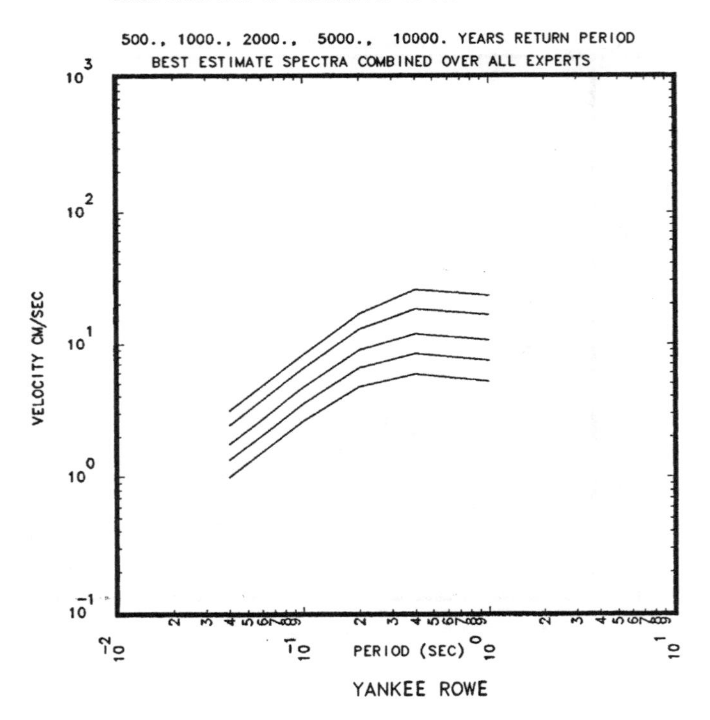

Figure 2.19.5 BEUHS for return periods of 500, 1000, 2000, 5000 and 10000 years aggregated over all S and G-Experts for the Yankee at Rowe site.

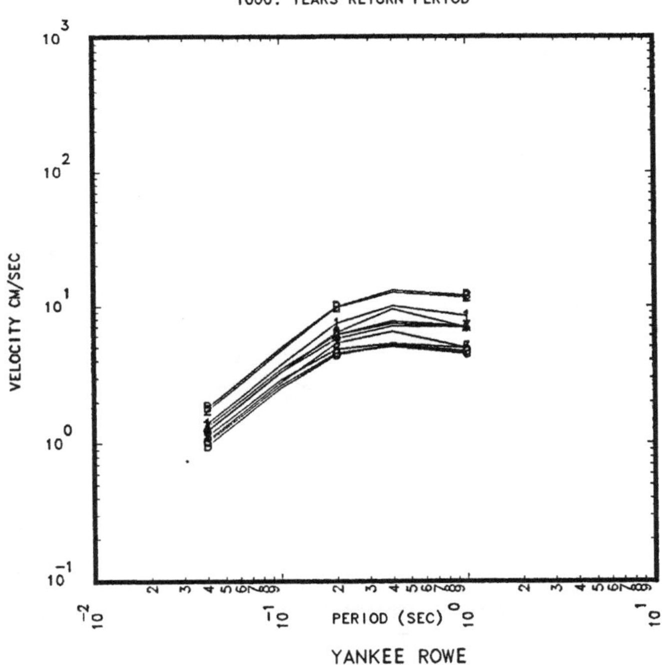

Figure 2.19.6 The 1000 year return period BEUHS per S-Expert aggregated over all G-Experts for the Yankee at Rowe site. Plot symbols are given in Table 2.0.

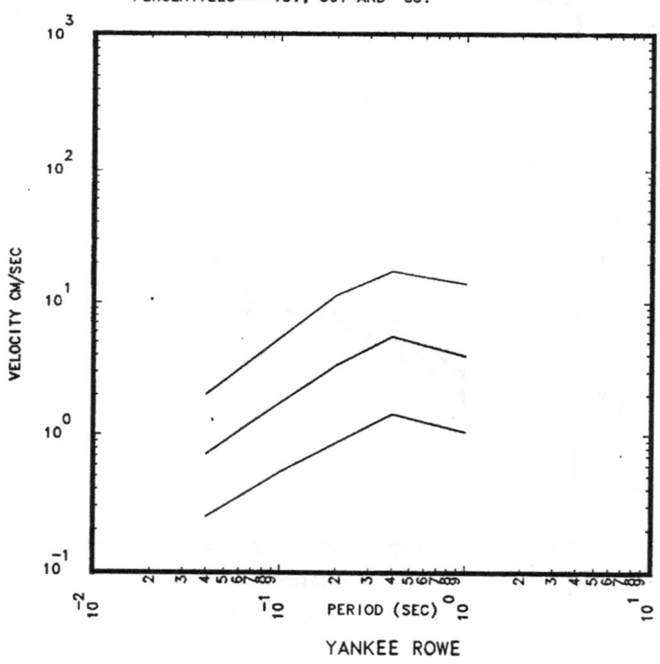

Figure 2.19.7 500 year return period CPUHS for the 15th, 50th and 85th percentiles aggregated over all S and G-Experts for the Yankee at Rowe site.

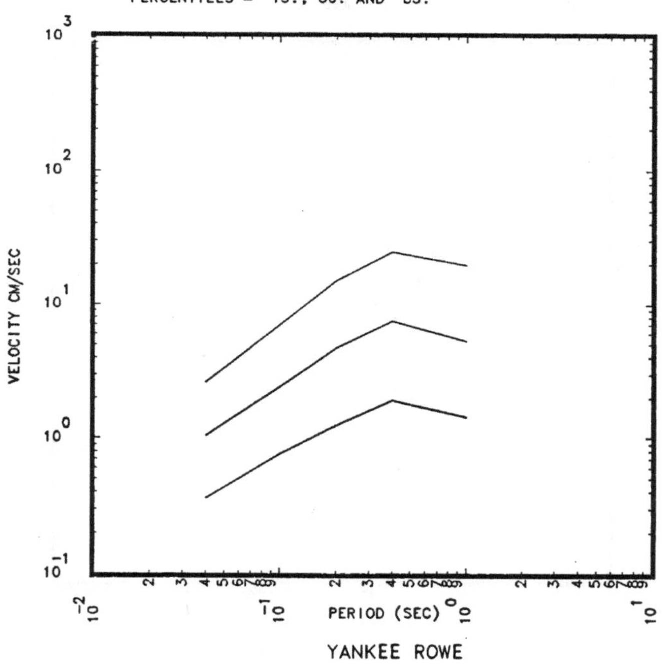

Figure 2.19.8 1000 year return period CPUHS for the 15th, 50th and 85th percentile aggregated over all S and G-Experts for the Yankee at Rowe site.

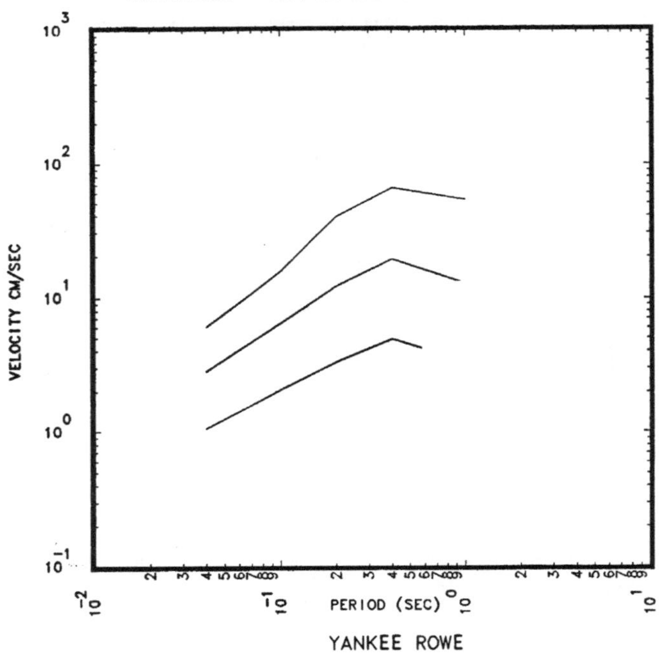

Figure 2.19.9 10000 year return period CPUHS for the 15th, 50th and 85th percentiles aggregated over all S and G-Experts for the Yankee at Rowe site.

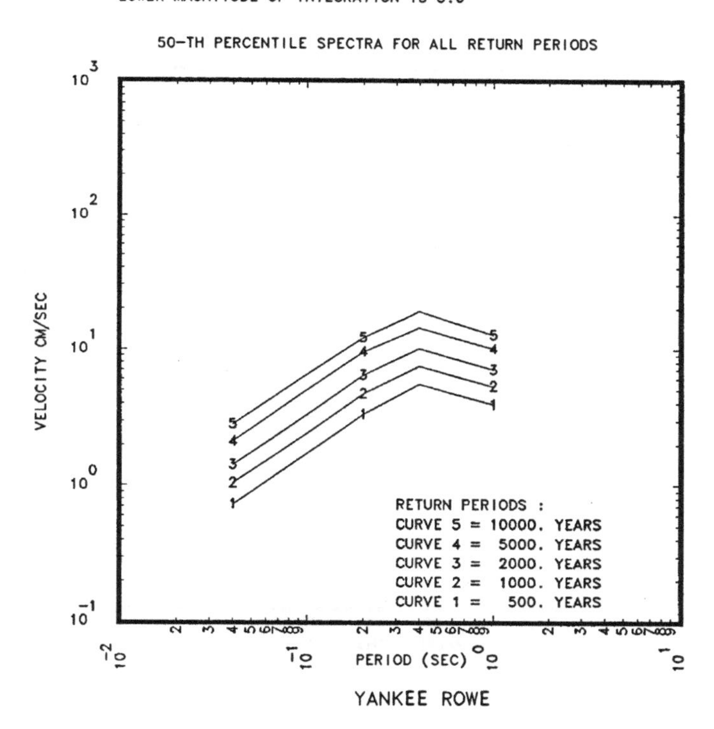

Figure 2.19.10 Comparison of the 50th percentile CPUHS for return periods of 500, 1000, 2000, 5000 and 10000 years for the Yankee at Rowe site.

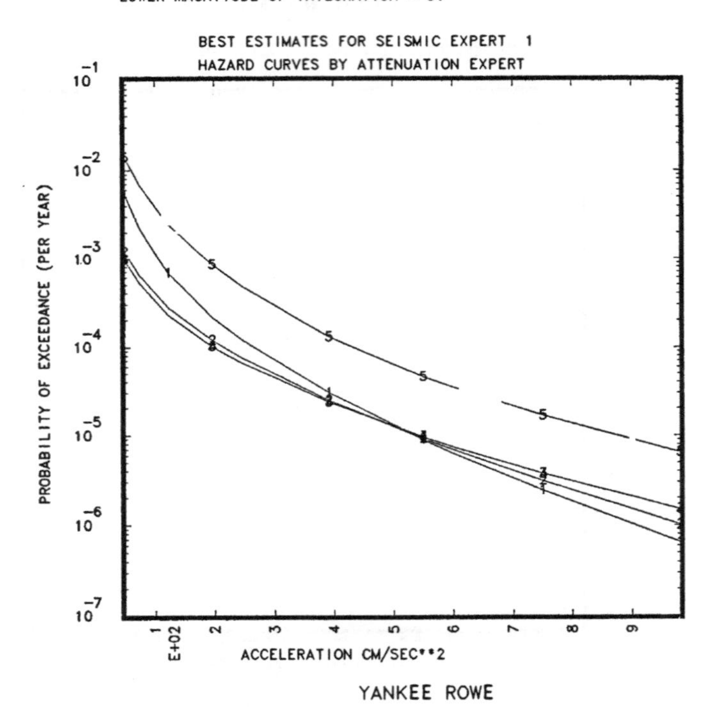

Figure 2.19.11. Comparison of BEHCs for PGA per G-Expert for S-Expert 1's input for the Yankee Rowe site. The spread in BEHCs between G-Expert 5's and the other G-Experts' BEHCs is typical for S-Experts 1, 2, 6, 7, and 13 for the Yankee Rowe site.

3. GENERAL DISCUSSION, REGIONAL OBSERVATIONS, AND COMPARISONS BETWEEN SITES

3.1 Uncertainty

Section 2 shows that there are significant differences in the estimated seismic hazard for any site between various experts. In addition, each expert has typically expressed significant uncertainty about his own input. This uncertainty in turn leads to significant uncertainty in the aggregated estimate of the seismic hazard at any site. As explained in Volume 1, we have used a Monte Carlo analysis to develop the distribution function for the seismic hazard at any site. In addition to the uncertainty in the probability distribution due to the variety of experts there is some small variability in the results due to the limitations on the number of simulations in our Monte Carlo uncertainty analysis to reasonably define the various estimators of the seismic hazard at a site. For example, in Fig. 3.1.1a we compare the CPHCs for PGA between Hope Creek and Salem sites and in Fig. 3.1.1b the 1000 year return period CPUHS for the same two sites. Hope Creek and Salem are located side by side, as can be seen from Fig. 1.1. Thus the comparison represent two different sets of simulations for the same site. We see from Figs. 3.1.1a,b that the median curve is well established and that there are only slight differences between the bounds (as measured by the 15th and 85th percentile curves). There are also only slight differences between the AMHCs for the two different Monte Carlo runs.

A detailed examination of the 2750 hazard curves generated in a Monte Carlo run for a typical site suggests that the distribution is approximately lognormal. However there are some sites where there is a marked departure from a lognormal distribution such as for the Fitzpatrick site. In Fig. 3.1.2a we compare the CPHCs for PGA for two different Monte Carlo runs and in Fig. 3.1.2b we compare the 1000 yr return period CPUHS for two different Monte Carlo runs both for the Fitzpatrick site. We see from Figs. 3.1.2a,b that there is a larger difference between the median curves for the two Monte Carlo runs than shown in Figs. 3.1.1a,b. In fact, Figs. 3.1.2a,b suggest that the 15th and 85th percentile curves are better defined than the median. If reference is made to Figs. 2.1.11 and 2.1.12 it is easy to see what is occurring. That is, as discussed in Section 2.1, G-Expert 5's GM model leads to hazard curves significantly higher than for most other GM modes. Thus, the distribution of hazard curves departs significantly from a lognormal distribution and becomes somewhat bimodal so that the 15th and 85th percentile curves are relatively well defined but the 50th percentile curve is somewhat "poorly" defined. However, even in this worst case, the differences between two Monte Carlo runs are not really large. Nevertheless, these differences between two Monte Carlo runs must be kept in mind when making comparisons between nearby sites.

3.2 Sensitivity to Region Choice

In Volume I we indicated that we approximately divided the EUS into four regions in order to give our G-Experts a chance to introduce regional corrections for attenuation and to give our S-Experts a chance to account for varying expertise for different regions. The boundaries between these regions are very approximate, thus as indicated in Section 2, all Batch 1 sites were considered to be located in region 1. However, several sites could be also considered to be in either region 2 or 3. The major impact of region placement is due to G-Expert 2's input. As can be seen from Tables 3.5 and 6 of Volume 1, only G-Expert 2 introduced a regional variation in his GM models. In region 1 he selected a different BE model than for regions 2,3 and 4. Thus the BEHC and BEUHS change depending upon whether the site is considered to be Region 1 or the other regions. In addition, it depends on whether it is a rock or a soil site with the effect being larger for a soil site than for a rock site. In Fig. 3.2.1 we show a comparison between the PGA hazard curves for the case when the Hope Creek site is considered to be located in region 1 and the case when it is considered to be located in region 2. We see from Fig. 3.2.1 that there is a noticeable difference between the BEHCs and the AMHCs for the two cases. There is however, some difference between the CPUHS as shown in Figs. 3.2.2 where 10,000 year return period CPUHS are plotted for the two cases.

The sensitivity to location placement is even less for rock sites as shown in Figs. 3.2.3 and 3.2.4. In Fig. 3.2.3 we show the BEHCs and AMHCs for the case when the Limerick site is considered to be located in region 1 and the case when it is considered to be located in region 2. It can be seen from Fig. 3.2.3 that the difference between the two cases is less than shown in Fig. 3.2.1. The difference between the CPHCs and the CPUHS for the two cases for Limerick site is small.

Overall we see that, for the sites located near the border between what we have defined as region 1 and the other regions, makes very little differences on the CPHCs and CPUHS whether we say they are located in region 1 or the other regions. It makes some differences on the BEHCs and AMHCs, particularly for soil sites. But given the overall uncertainty, as measured by the spread between the 15th and 85th percentile CPHCs the sensitivity is small but still larger than the variation due to the limitations in the number of Monte Carlo simulations.

3.3 Factors Influencing Zonal Contribution to the Hazard

A number of factors influence how significantly a given seismic source zone contributes to the hazard at any given site for any given S-Expert. Several factors are obvious and a few may not be so obvious. The main factors that influence zonal contribution to the hazard can be separated into three groups:

1. Attributes of the zone in question:

   - Distance from the zone to the site.
   - The rate of activity in the zone.
   - The b-value used for the zone.
   - The upper magnitude cut off for the zone.
   - The probability of existence of the zone.
   - The size of the zone.

2. Attributes of the ground motion model:

   - The rate at which the peak ground motion attenuates with distance.
   - The site's soil category.

3. Attributes of the hazard analysis methodology:

   - Uncertainty analysis performed.
   - Lower bound of integration for magnitude.

Let us start our discussion on the significance of the above factors in the inverse order, i.e. start with set listed under (3) above. We made the point in Section 2 that the Tables 2.SN.1 were based only on BE input and thus did not in all cases capture the true contribution of a given zone to the hazard for a specific site. The contribution of a given zone listed in Tables 2.SN.1 might be too high if a zone's probability of existence is relatively low but greater than or equal to 0.5. On the other hand it is too low (not listed) if the probability of existence is less than 0.5.

In our analysis the modeling uncertainty in the site correction is accounted for by allowing for several different type of corrections to be performed. The type of correction performed also can impact how the ground motion model (or models) impact the seismic hazard and how correction for the site's soil category is made. This will be discussed later. Other elements of the uncertainty analysis such as the variation in zone boundary shape, variation in rate of activity etc. are less important and generally do not play a major role in determining the zonal contribution to the hazard.

The lower bound for magnitude used in the analysis is of some significance at the low g-value end. This is illustrated in Figs. 3.3.1a and b. In Fig. 3.3.1a we show the contribution to the BEHC for PGA from all the earthquakes in four distance rings about the Maine Yankee site when the lower bound of integration for magnitude is 5.0. Similarly in Fig. 3.3.1b we show the same thing except the lower bound of integration has been reduced to 3.75. We see by comparing Fig. 3.3.1a to b that, when the lower bound of integration is lowered to 3.75, the region from $0 \leq$ Distance ≤ 15 contributes significantly more to the hazard at lower PGA levels and there is also a marked increase from the region $15 <$ Distance < 50. At high PGA levels there is little effect of changing the lower bound of integration. Of course, if the lower bound was yet even higher than 5, the effect would be more significant even to much higher PGA levels.

Let us now address group (2) - attributes of the GM model. First, let us note if there was no modeling uncertainty, i.e., if we knew the correct form for the GM model then the attributes of the GM model would not influence the zonal contribution. If no uncertainty analysis is being performed e.g. Algermissen et. al. (1982) then it is the same as saying we know the correct form for the GM model. In our analysis we have included uncertainty about GM modeling in three ways:

(1) We used multiple GM models.
(2) We introduced multiple ways to correct for the effect that the site's category has on the estimated ground motion.
(3) We varied the random uncertainty associated with each GM model.

All three of the above are important.

In section 2.1 we have already made the point that one of the BE GM models has a significantly lower attenuation rate than the other BE GM models. This can make a significant difference as is illustrated in Figs. 3.3.2a and b. In Fig. 3.3.2a we show the contribution to the BEHC for PGA from the earthquakes located in four distance ranges when all the BE GM models are used for the Fitzpatrick site. In Fig. 3.3.2b we show the same thing as in Fig. 3.3.2a except only G-Experts 1-4's BE GM models are used. That is, we have eliminated the GM model with low attenuation. It is evident from comparing Figs. 3.3.2a and b that the uncertainty about the correct GM model is very significant and can have an important impact on determining which zones contribute to the hazard at a site.

The site's soil category has an influence because our uncertainty about how to correct the GM for the soil conditions at a site impacts the estimate of the ground motion at a site from an earthquake differently for various ground motion models. This is illustrated in Figs. 3.3.3a and b. In Fig. 3.3.3a we show the contribution to the BEHC for PGA for four distance ranges for the Vermont Yankee site. The Vermont Yankee site is a rock site. In Fig. 3.3.3.b we show the contribution to the BEHC for PGA for the same four distance ranges for the Yankee Rowe site. The Yankee Rowe site was placed in the Till-2 soil category. We see from Fig. 1.1 that Yankee Rowe (plot symbol J) and Vermont Yankee (plot symbol I) are located relatively near each other, thus we would expect little difference in the seismic hazard between these two sites. We see however from comparing Figs. 3.3.3a and b that there is a considerable difference between the two sites in the distance ranges which contribute most to the BEHCs for PGA. Thus if Table 2.18.1 is compared to Table 2.19.1 we see some significant differences. Most of these differences arise from the differences in the estimates of the PGA for a given earthquake at the two sites due to the correction for site soil conditions.

It should be noted that the differences between the distribution of which distance bands contribute most to the BEHCs for PGA for rock sites as compared to soil sites is typically the difference between Figs. 3.3.3a and b. However, there is some variation between the various rock sites and the various soil sites depending upon their location. For example, by comparing Figs. 3.3.4a and b to Figs 3.3.3a and b we see that there is a difference in the distribution of the distance ranges which contribute most to the BEHCs for PGA for sites located in the southern part of region 1, such as Limerick (rock) and Salem (soil), and sites located in the northern part of region 1 such as Vermont Yankee and Yankee Rowe.

The group (1) - attributes of the zone in question-are relatively easy to understand and are generally the factors which we expect to control a given zone's contribution to the seismic hazard at a site. In Section 2 we gave examples of how the group (1) factors influence this contribution.

From the above discussion we can conclude that care must be taken when using the information given in Table 2.1.1 to 2.19.1. The information is useful, but, as indicated, it can give a distorted picture of which zones are most significant. Unfortunately, in complex cases such as the Fitzpatrick site the only way to get an undistorted understanding is to perform a detailed sensitivity analysis. However, one can gain a relatively good understanding of what is important by carefully examining the data given in Tables 2.SN.1, the zonation maps for each S-Expert, the seismicity data for each S-Expert given in Vol. I Appendix B and keeping in mind the sensitivities discussed in this section.

3.4 Comparisons of the Seismic Hazard Between Sites

In this project the seismic hazard has been defined as the annual probability of exceedance of a given level of peak ground motion. Thus, strictly speaking, we only need to compare hazard curves between sites to reach a conclusion about the relative hazard at various sites. However, given the large uncertainties that exist in our estimate of the peak ground motion and our inability to convert a given level of peak ground motion into a risk number, suggests that we need to also introduce some subjective judgement into the process of assessment of the relative hazard between sites. In this section we compare the computed seismic hazard between sites. In addition we examine some important elements that should be factored into an assessment of the relative difference in the seismic hazard between the sites located in region 1.

When comparing the hazard between two sites, one of the most important factors to be considered is the potential difference in soil categories between the two sites. For example, in Fig. 3.4.1 we compare the median CPHC for the Vermont Yankee site to the median CPHC for the Yankee Rowe site. We see from Fig. 1.1 that these two sites are nearby each other hence we would generally expect that there would be little difference between the seismic hazard for these two sites. However, as can be seen from Fig. 3.4.1, there is considerable difference in the hazard between these two sites. The reason for this difference is the fact that Vermont Yankee is a rock site and Yankee Rowe is a soil site (Category Till-2). A complete discussion on the local site effects on ground motion is given in Volume VI, however, Fig. 3.4.1 shows that it is very significant. The uncertainty surrounding the correction for local site effects makes it difficult to unequivocally conclude that the seismic hazard is higher at the Yankee Rowe site than at the Vermont Yankee site as the amplification appearing in the final results is a result of the very complex interaction of the ground motion models, the actual correction factor and the magnitude and distance of the earthquakes.

In addition to soil category, one must consider which estimator of the hazard should be used because in some cases different estimators would lead to different conclusions about the relative hazard between two sites. For example, if Fig. 2.18.1 is compared to Fig. 2.19.1 we see that both the AMHC and BEHC are higher at the Vermont Yankee site than at the Yankee Rowe site yet the median CPHC is significantly higher at the Yankee Rowe site than at the Vermont Yankee site. If Fig. 2.18.3 is compared to Fig. 2.19.3 we see that there is little difference between the 15th CPHCs and little difference between the 85th CPHCs at the two sites. The main reason the BEHC and AMHC are so "high" at the Vermont Yankee site as compared to Yankee Rowe can be seen by comparing Fig. 2.18.11 to Fig. 2.19.11 which gives the spread between BEHCs per G-Expert for a particular S-Expert (different S-Experts are shown here but each are typical for these two sites). We see for Vermont Yankee that one G-Expert is much higher than the others, where as at Yankee Rowe the BEHCs for all G-Experts are about the same. The high outliers tend to raise both the BEHC and AMHC at the Vermont Yankee site.

Most of the comparisons made in the rest of this section will be relative to the median hazard curve because it proved to be the most stable parameter among all the parameters displayed in this report (i.e., BEHC, median and AMHC). Generally, the spread between the 15th and 85the CPHCs gives an idea of uncertainty at a given site, however, as we have seen at the Vermont Yankee site, this spread may not fully capture the uncertainty in which case there will generally be a larger spread between the 50th and 85th CPHCs than the 15th and 50th CPHCs.

As one would expect, the hazard is not uniform over region 1 as is shown in Fig. 3.4.2. In Fig. 3.4.2 we compare the median CPHCs for five rock sites spread across region 1. It can be seen from Fig. 3.4.2 that there is a wide variation in the seismic hazard in the region.

It is somewhat instructive to compare the contribution to the BEHCs for various magnitude and distance ranges between the five sites plotted in Fig. 3.4.2. The contribution to BEHCs from earthquakes in 4 magnitudes ranges are given in Figs. 2.1.4 for the Fitzpatrick site, Fig. 2.4.4 for the Limerick site, Fig. 2.7.4 for the Maine Yankee site, Fig. 2.8.4 for the Millstone site and Fig. 2.14.4 for the Seabrook site. Comparing these figures we see some differences between the sites. For example, larger magnitude earthquakes contribute more significantly to the hazard at the Maine Yankee site than at the other sites, and conversely, large magnitude earthquakes are less significant at the Limerick site than at the other sites.

Figure 3.4.3 gives the contribution to the BEHC for the Seabrook site from four distance ranges, Fig. 3.3.2c for the Fitzpatrick site, Fig. 3.3.4a for the Limerick site, and Fig. 3.3.1c for the Maine Yankee site. The relative contribution to the BEHC for the Millstone site for the four distance ranges is similar to the Seabrook site and hence, does not need to be repeated. There are some interesting differences between the relative contribution to the BEHC from each of the four distance ranges for several of the sites that are compared in Fig. 3.4.2. For example, we see at the Fitzpatrick site that most of the contribution to the BEHC for PGA is from earthquakes farther than 50 km away. This is certainly one reason why the hazard at the Fitzpatrick site is lower than at the other sites in region 1. Distant earthquakes are also somewhat more important at the Maine Yankee site than at the Seabrook, Millstone and Limerick sites. Distant earthquakes are least important at the Limerick site. This latter fact also goes with the fact that large earthquakes are least important at the Limerick site.

In Fig. 3.4.4 we compare the median CPHCs for the Hope Creek (deep soil), Pilgrim (sand-2), Shoreham (deep soil) and Yankee Rowe sites (till-2). This comparison is less useful than the one made in Fig. 3.4.2 for rock sites because there are three different soil categories involved. Only Shoreham and Hope Creek have the same soil category.

It is evident that there can be some variation in the seismic hazard over a relatively small area. This is illustrated in Figs. 3.4.5a, b,c. In Fig. 3.4.5a we compare the median CPHCs for the Seabrook, Maine Yankee and Vermont Yankee sites. It should be noted that the median CPHC for the Millstone and Haddam Neck sites is similar to the Vermont Yankee site, however the make-up of the factors (e.g., the magnitude and distance ranges) which contribute to the hazard are different between Vermont Yankee, Millstone and Haddam Neck. In Fig. 3.4.5b we compare the median CPHCs for the Hope Creek, Oyster Creek and Shoreham sites. In Fig. 3.4.5.c we compare the median CPHCs between the Limerick and Susquehanna sites. It should be noted that the median CPHC at the Peach Bottom and Three Mile Island sites is more similar to the Limerick site than to the Susquehanna site.

The spectra curves show much the same differences between sites as the PGA curves. However because the spectral ordinates are plotted on a log scale whereas the PGA is plotted on linear scale, the differences between sites are more noticeable for PGA comparisons than for spectral comparisons. Nevertheless some interesting differences can be observed by making spectral comparisons. For example, in Fig. 3.4.6a we compare the 10,000 year return period median CPUHS for the Limerick and Millstone sites. We pointed out earlier that smaller nearby earthquakes are more significant at the Limerick site than at the Millstone site. The shapes of the CPUHS for the two sites reflect this. We see from Fig. 3.4.6a that the CPUHS for the Limerick site have more short period energy and less long period energy at the Millstone site. This effect can be seen also in Fig. 3.4.6b where we compare the 10,000 year return period median CPUHS for the Seabrook, Maine Yankee, Millstone and Fitzpatrick sites. We pointed out earlier that, relatively speaking, large earthquakes were significant at the Maine Yankee site than at the other sites shown in Fig. 3.4.6b and least important at the Millstone site. The shape of the CPUHS curves reflect this effect at the long period end of the spectrum. Additional comparisons and general conclusions are given in Volume VI.

In Fig. 3.4.7 we compare the median CPHC for all the sites listed in Table 1.1. The variation between sites includes all the factors we have discussed, such as site category and differences in zonation between sites. Figure 3.4.7 shows that the four sites north of the highest probability of exceedance of 0.2g are sets number 12(C), 14(E), 7 and 19(J), (i.e., Pilgrim, Seabrook, Maine Yankee and Yankee Rowe).

It should be kept in mind that much of the above was based on comparisons between median estimators of the hazard at various sites. As we have pointed out earlier, other estimators of the hazard might lead to different conclusions. To illustrate this point, refer to Fig. 3.4.8, according to which, the four sites with the highest probability of exceedance of 0.2g are sites number 2, 7, 12 and 14 (i.e., Ginna, Maine Yankee, Pilgrim and Seabrook) when the estimator is the arithmetic mean (symbol "A" on Fig. 3.4.8). When the estimator is the median (symbol "M"), the four sites are 7, 12, 14 and 19 (Yankee Rowe). When the estimator is the best estimate (symbol "B"), the four sites are 2, 7, 9 (Nine Mile Point) and 14, and finally when the estimator is the 85th percentile, the four sites are 7, 12, 14 and 19.

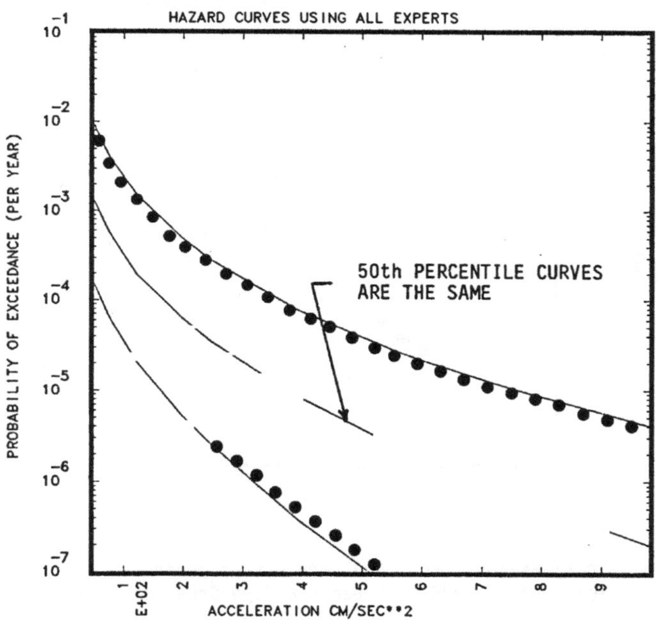

Figure 3.1.1a Comparison of the 15th, 50th and 85th percentile CPHCs for PGA between the Hope Creek and Salem sites.

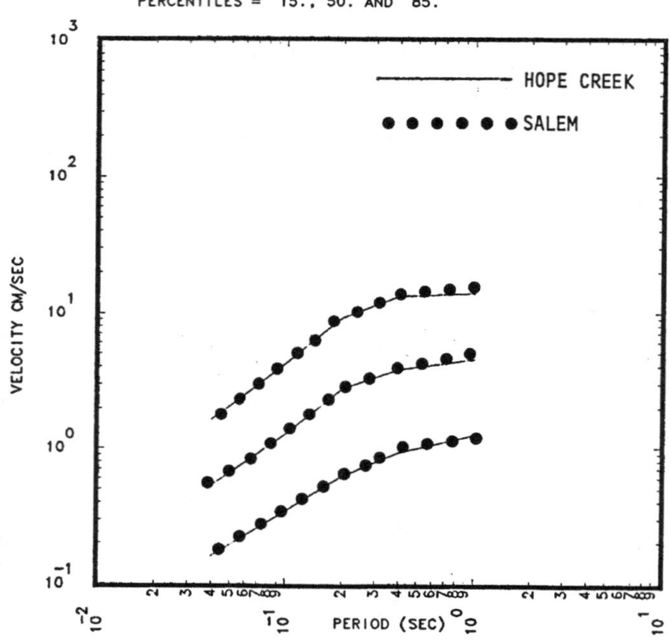

Figure 3.1.1b Comparison of the 1000 year return period 15th, 50th and 85th percentile CPUHS between the Hope Creek and Salem sites.

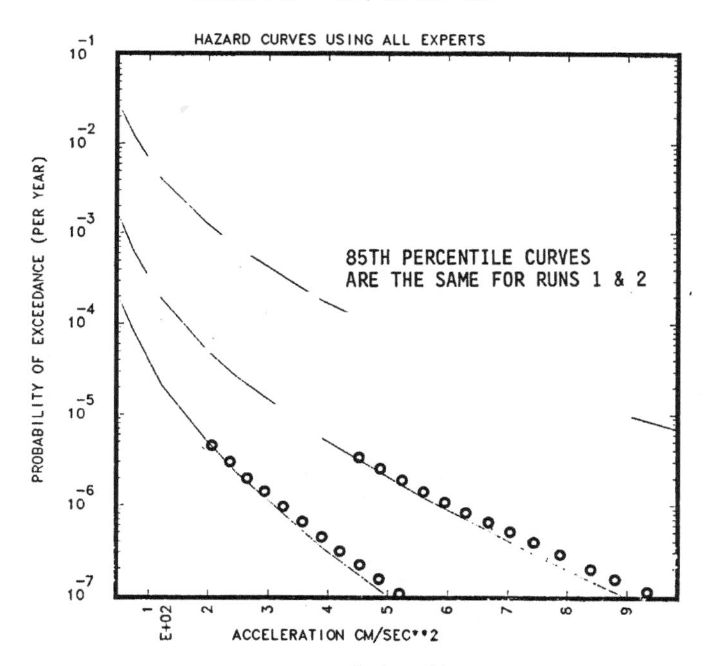

Figure 3.1.2a Comparison of the 15th, 50th and 85th percentile CPHCs for PGA between two different Monte Carlo runs for the Fitzpatrick site.

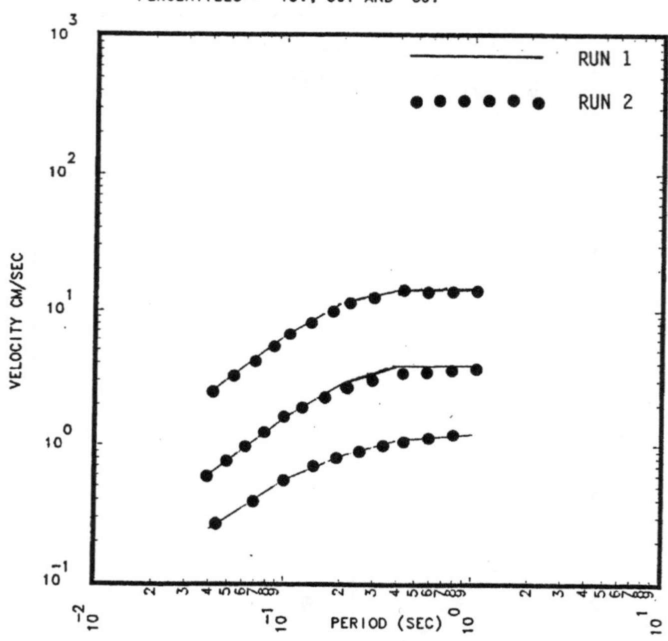

Figure 3.1.2b Comparison of the 1000 year return period 15th, 50th and 85th percentile CPUHS between two different Monte Carlo runs for the Fitzpatrick site.

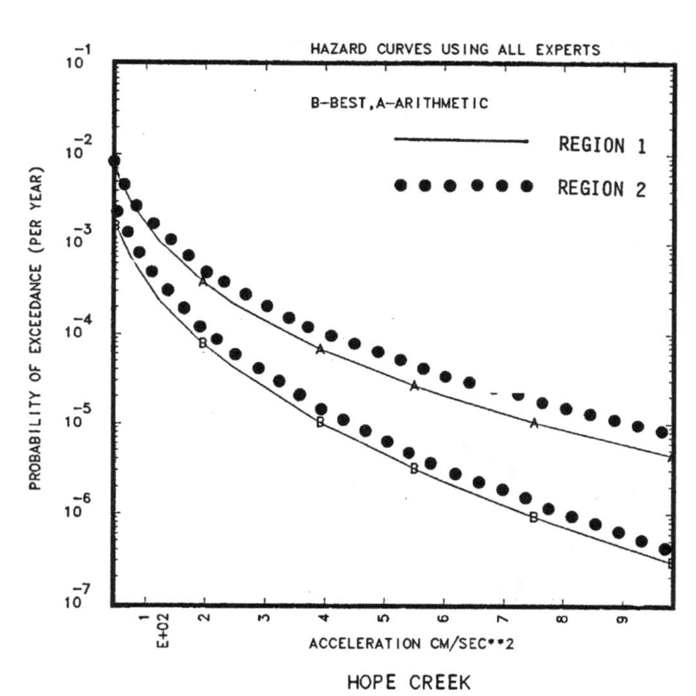

Figure 3.2.1 Comparison of the BEHCs and AMHCs between the case when the Hope Creek site is considered to be located in region 1 and the case when it is considered to be located in region 2.

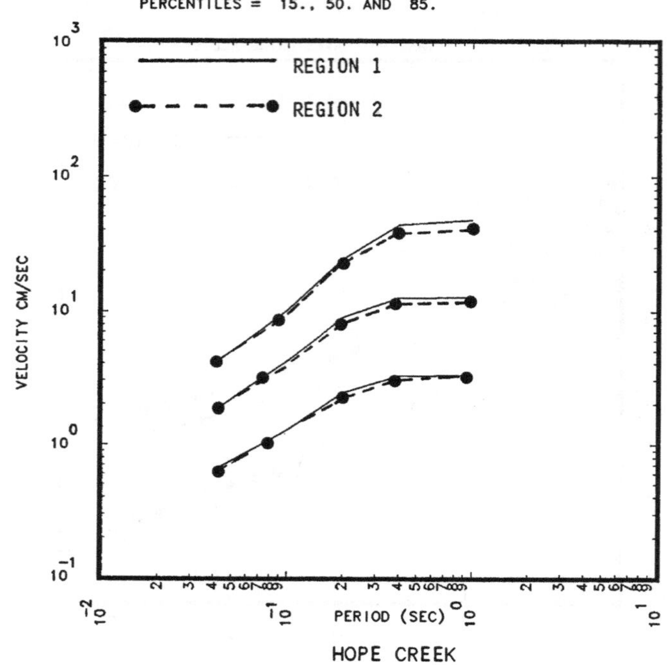

Figure 3.2.2 Comparison of the 10,000 year return period 15th, 50th and 85th percentile CPUHS for the Hope Creek site between the case when the Hope Creek site is considered to be located in region 1 and the case when it is considered to be located in region 2.

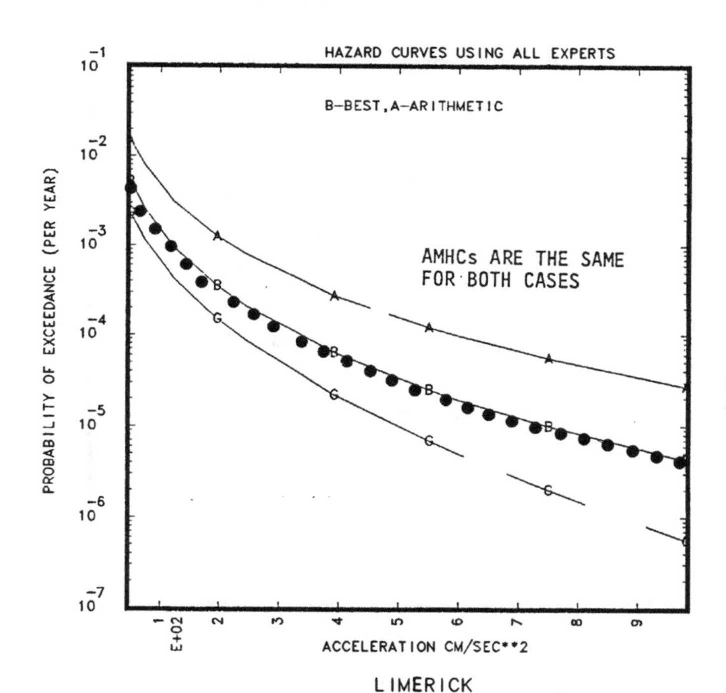

LIMERICK

Figure 3.2.3 Comparison of the BEHCs, AMHCs for the Limerick site between the case when the Limerick site is considered to be located in region 1 and the case when it is considered to be located in region 2.

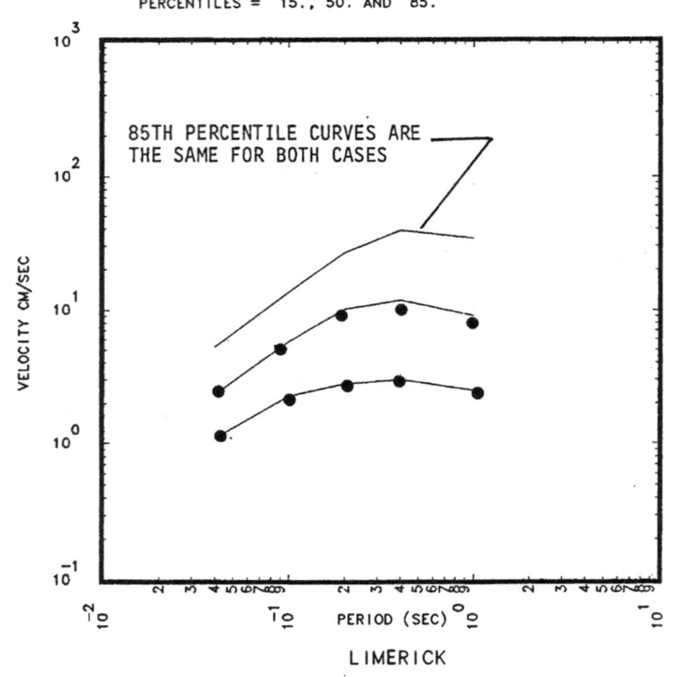

Figure 3.2.4 Comparison of the 10,000 year return period 15th, 50th and 85th percentile CPUHS for the Limerick site between the case when the Limerick site is considered to be located in region 1 and the case when it is considered to be located in region 2.

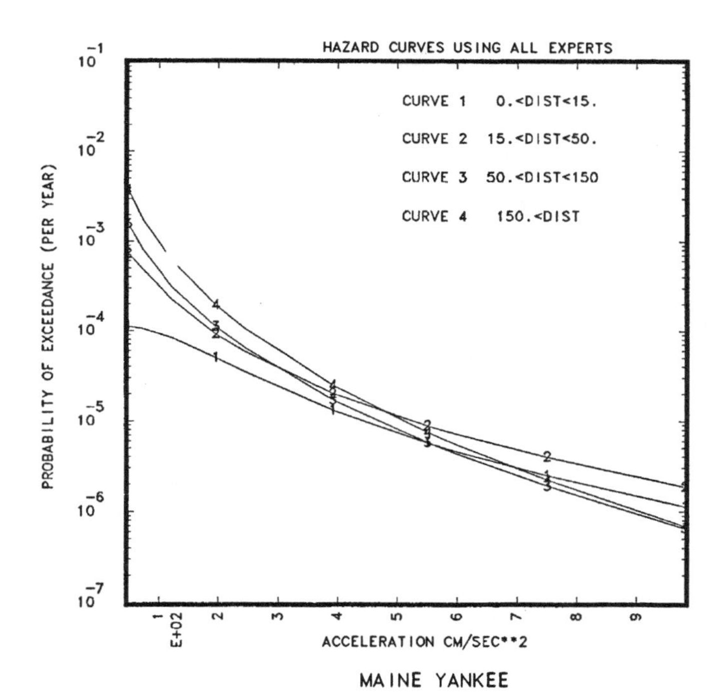

Figure 3.3.1a BEHCs which include only the contribution to the PGA hazard from earthquakes within the indicated distance ranges for the Maine Yankee site when the lower bound of integration for magnitude is 5.0.

CONTRIBUTION TO THE HAZARD FOR PGA
FROM THE EARTHQUKES IN 4 DISTANCE RANGES MO=3.75

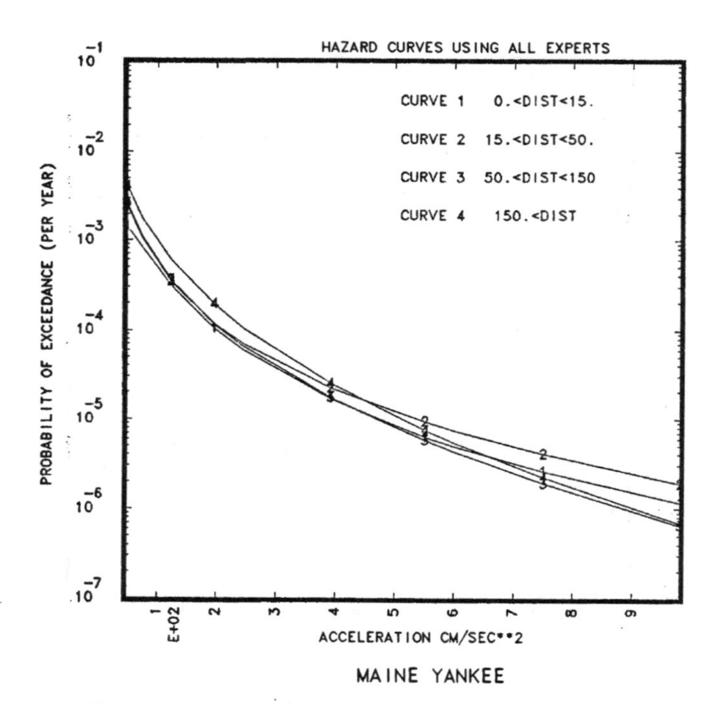

Figure 3.3.1b BEHCs which include only the contribution to the PGA hazard from earthquakes within the indicated distance ranges for the Maine Yankee site when the lower bound of integration is 3.75.

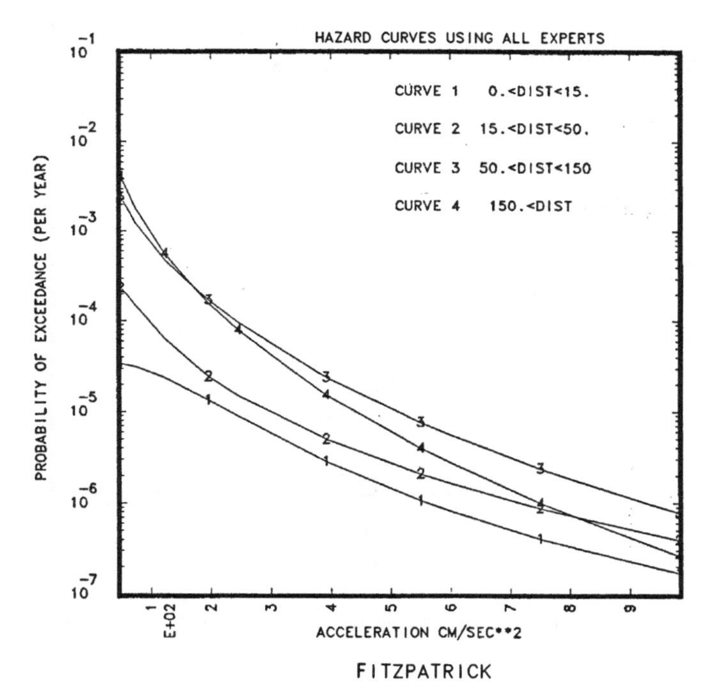

Figure 3.3.2a BEHCs which include only the contribution to the PGA hazard from earthquakes within the indicated distance ranges for the Fitzpatrick site.

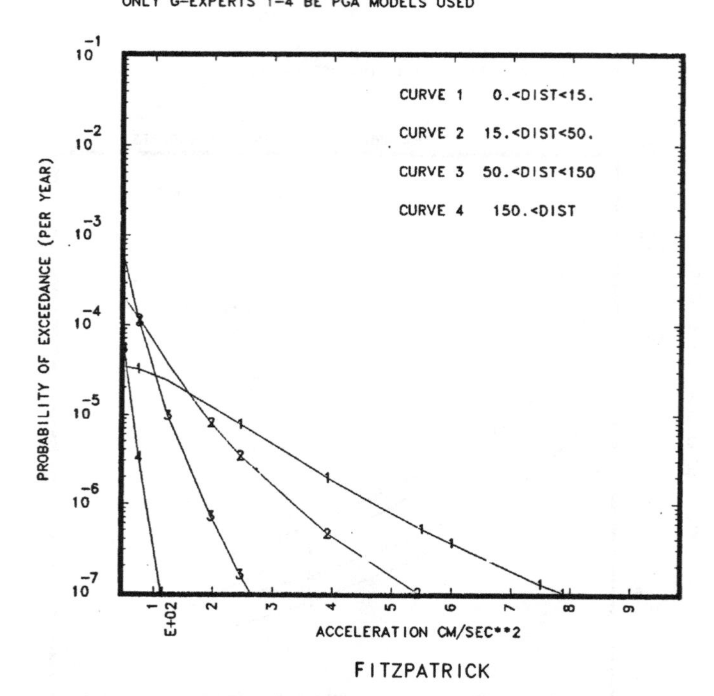

Figure 3.3.2b Same as Fig. 3.3.2a except only G-Experts 1-4 BE GM models were used.

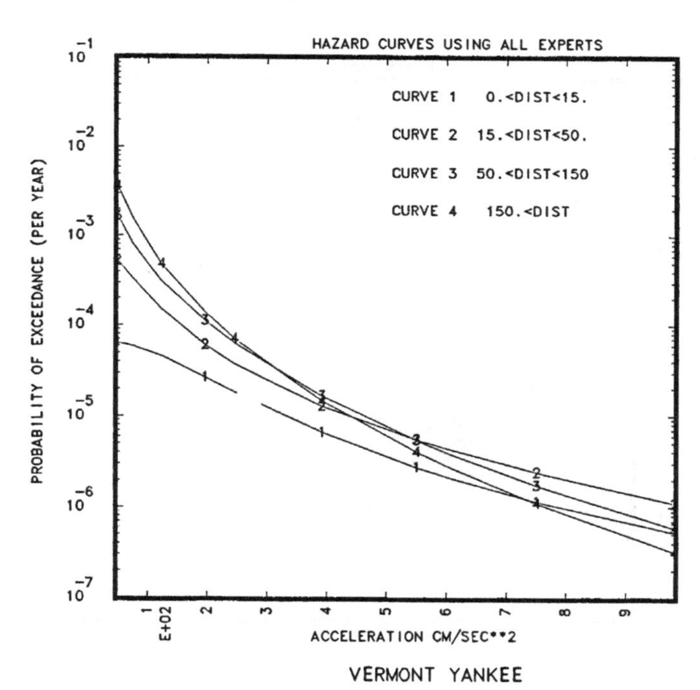

Figure 3.3.3a BEHCs for a rock site located in the northern part of region 1 which include only the contribution to the hazard for PGA from earthquakes within the indicated distance ranges.

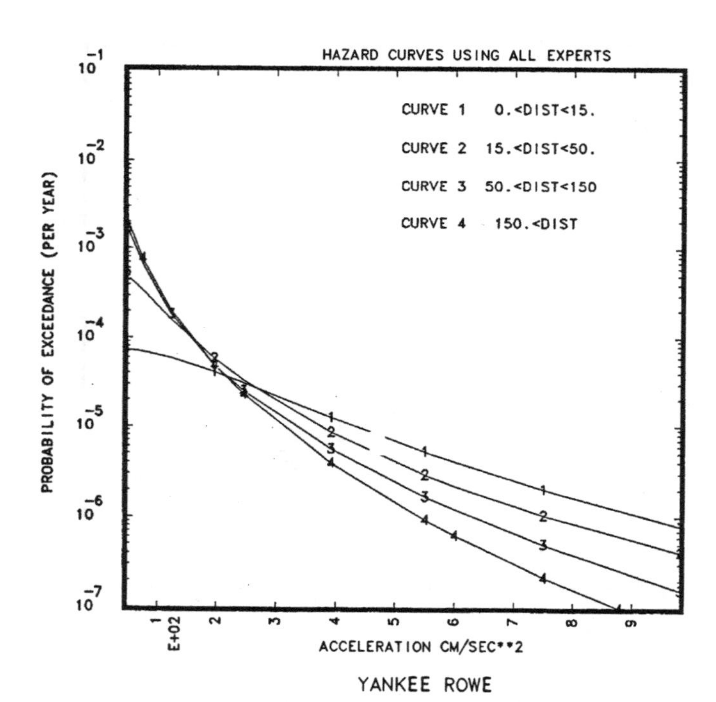

Figure 3.3.3b BEHCs for a soil site located in the northern part of region 1 which include only the contribution to the PGA hazard from the earthquakes within the indicated distance ranges.

CONTRIBUTION TO THE HAZARD FOR PGA
FROM THE EARTHQUKES IN 4 DISTANCE RANGES

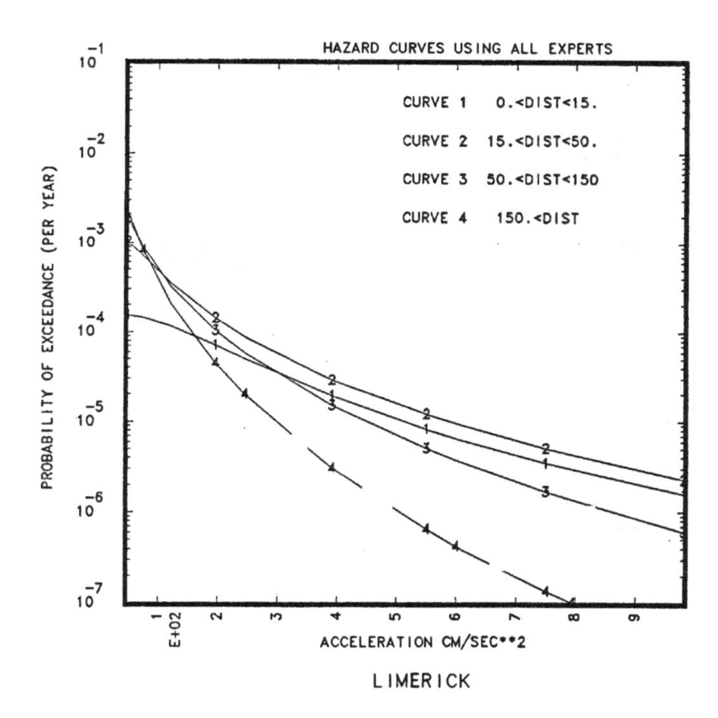

Figure 3.3.4a BEHCs for a rock site located in the southern part of region 1 which include only the contribution to the hazard for PGA from earthquakes within the indicated distance ranges.

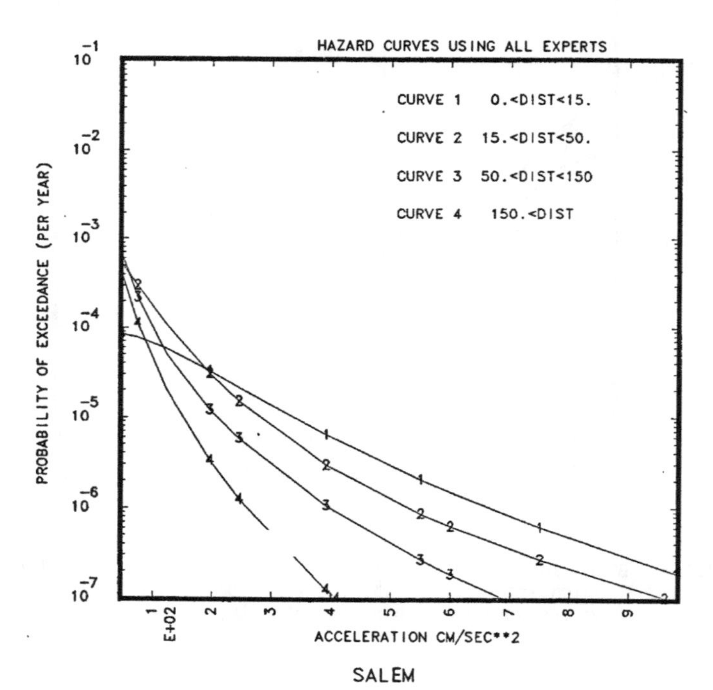

SALEM

Figure 3.3.4b BEHCs for a soil site located in the southern part of region 1 which include only the contribution to the hazard for PGA from earthquakes within the indicated distance ranges.

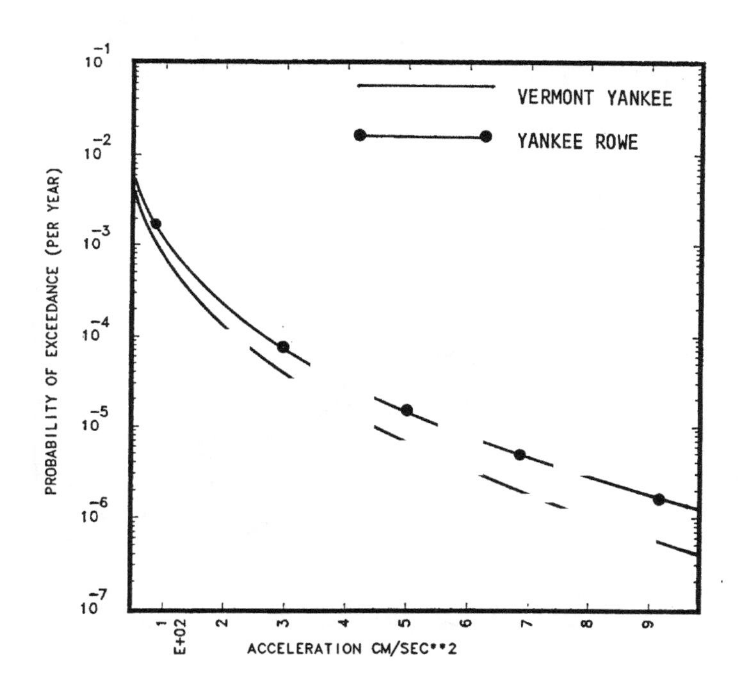

Figure 3.4.1 Comparison of the median CPHCs for PGA between the Vermont Yankee and Yankee Rowe site.

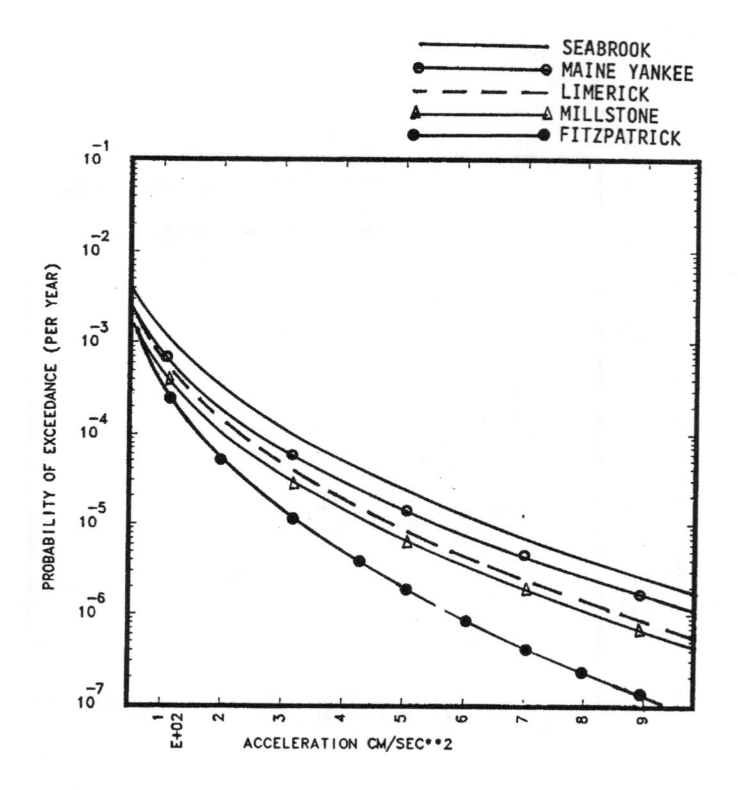

Figure 3.4.2 Comparison of the median CPHCs for PGA for the five sites indicated.

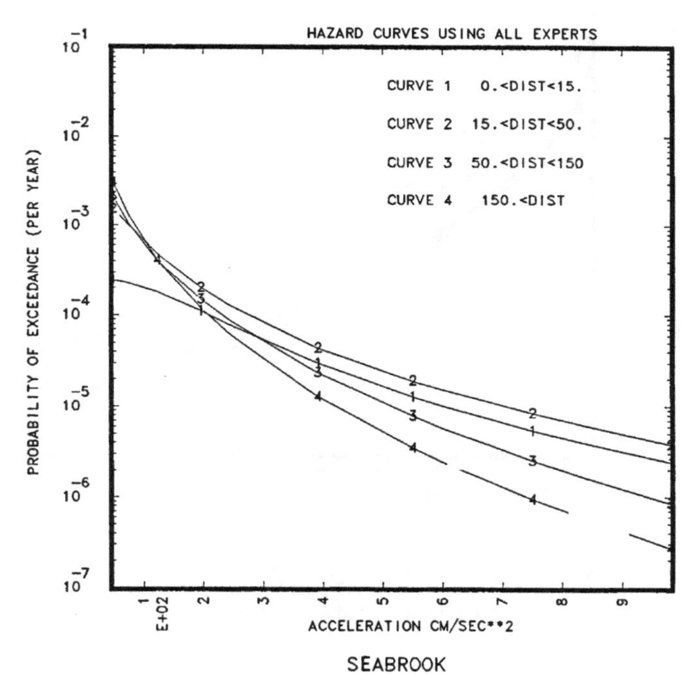

Figure 3.4.3 BEHCs which include only the contribution to the PGA hazard from earthquakes within the indicated distance ranges for the Seabrook site.

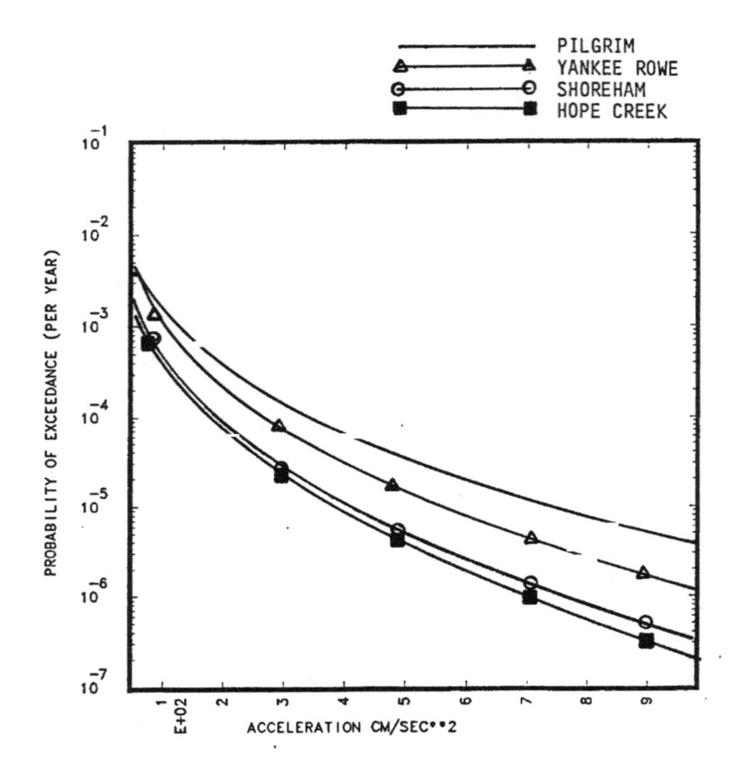

Figure 3.4.4 Comparison of the median CPHCs for PGA between the four soil sites indicated. It should be noted that the soil sites fall into several different soil categories.

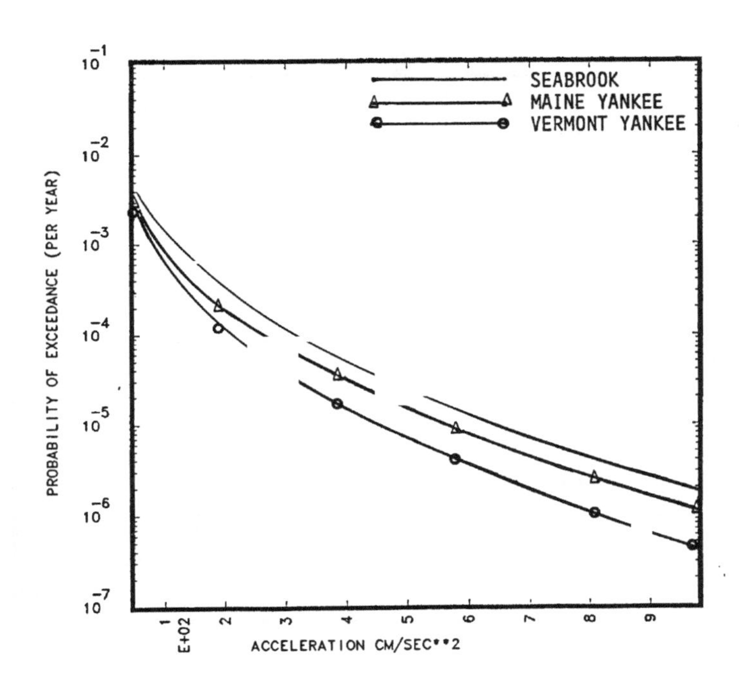

Figure 3.4.5a Comparison of the median CPHCs for PGA between the three nearby sites indicated.

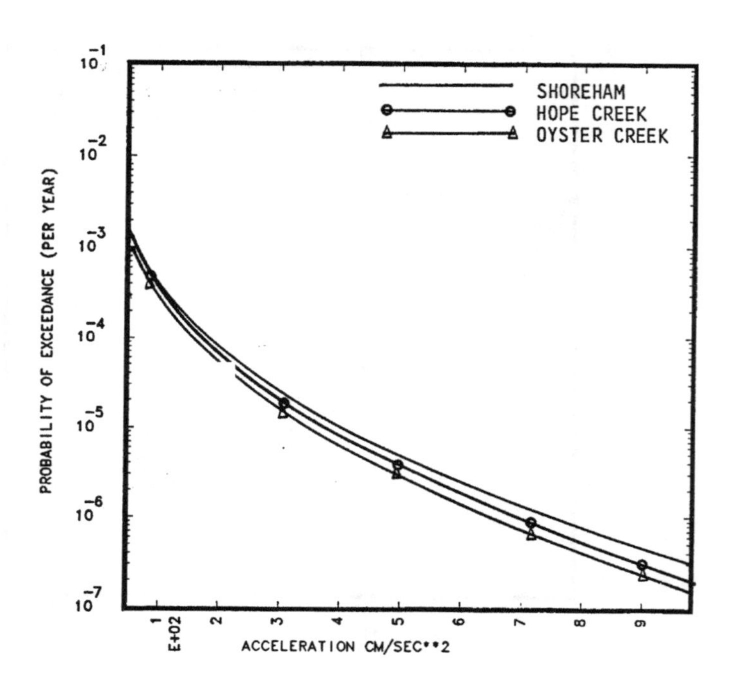

Figure 3.4.5b Comparison of the median CPHCs for PGA between the three nearby sites indicated.

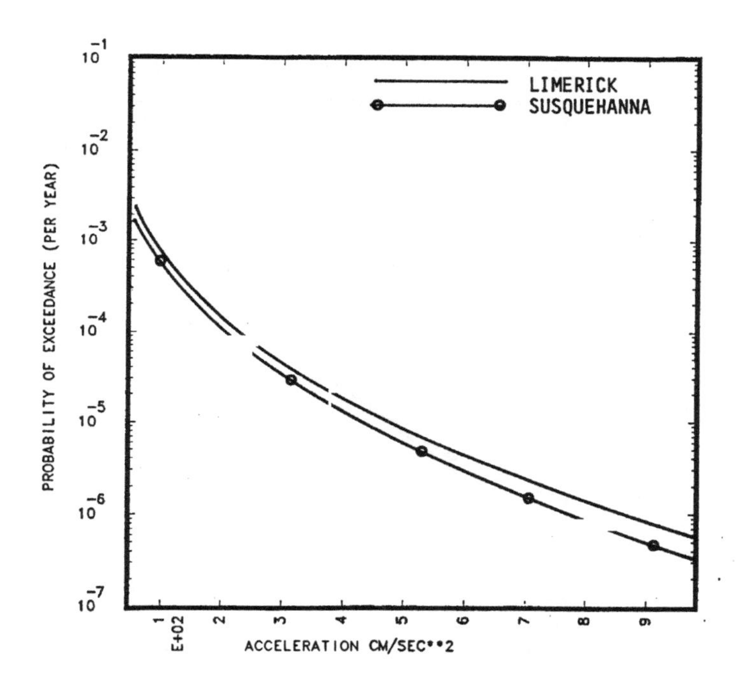

Figure 3.4.5c Comparison of the median CPHCs for PGA between the Limerick and Susquehanna sites.

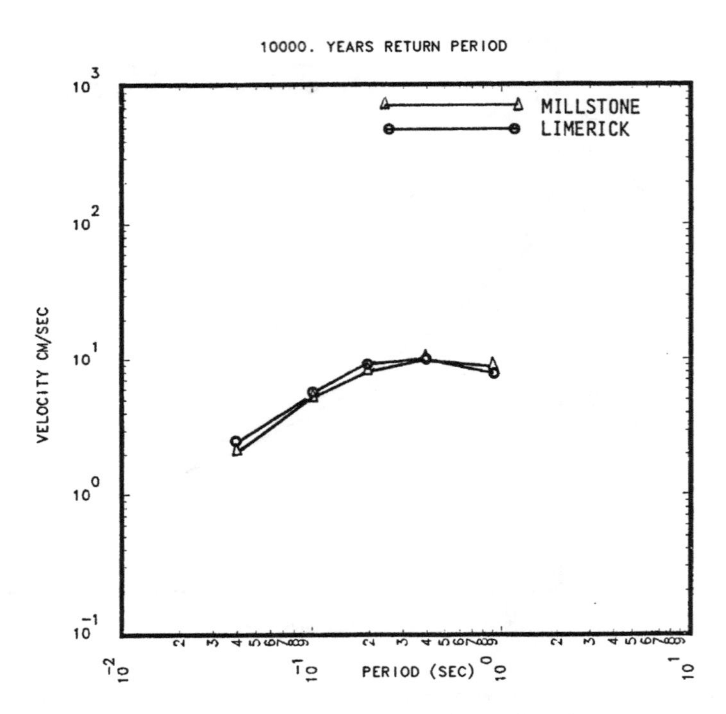

Figure 3.4.6a Comparison of the median 10,000 year return period CPUHS between the Limerick and Millstone sites.

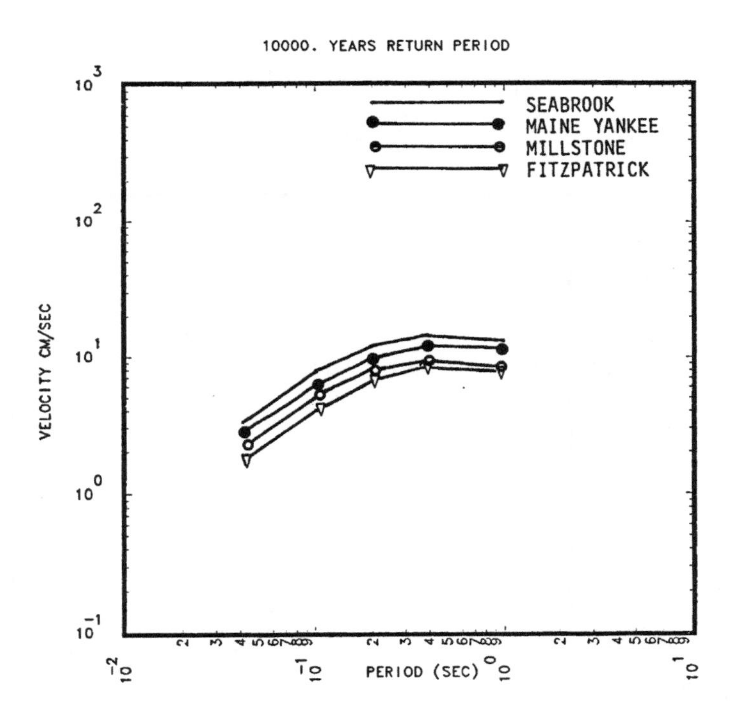

Figure 3.4.6b Comparison of the median 10,000 year return period CPUHS between the four sites indicated.

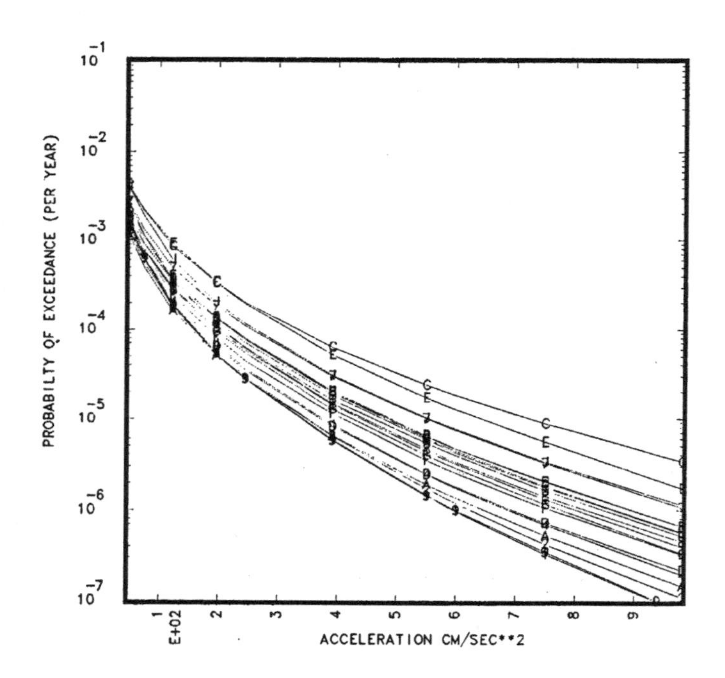

Figure 3.4.7 Comparison of the median CPHCs for the sites in Vol. II. The plot symbols used to identify the sites are given in Table 1.1.

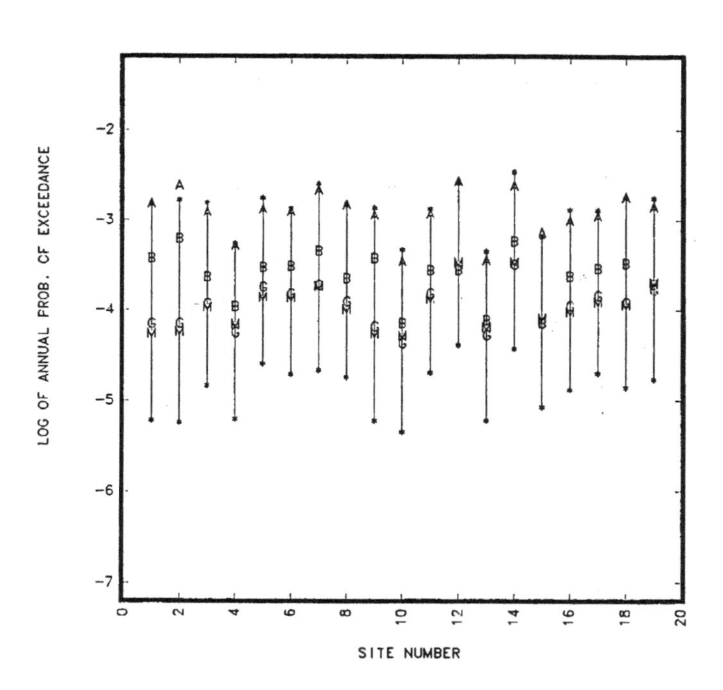

Figure. 3.4.8 Median (M) Probability of exceedance of 0.2g. Best estimate (B), Arithmetic mean (A) 15th and 85th percentiles (*) for the 19 sites of batch 1.

Appendix A

References

Algermissen, S.T., Perkins, D.M., Thenhaus, P.C., Hangen, S.L., and Bender, B.L.k (1982), Probabilistic Estimates of Maximum Acceleration and Velocity in Rock in the Contiguous United States, USGS, open file report 821033.

Bernreuter, D.L. and Minichino, C. (1983), Seismic Hazard Analysis Overview and Executive Summary, NUREG/CR-1582, Vol. 1 (UCRL-53030).

Bernreuter, D.L., Savy, J.B., and Mensing, R.W. (1987), Seismic Hazard Characterization of the Eastern United States: Comparative Evaluation of the LLNL and EPRI Studies, U.S. NRC Report NUREG/CR-4885.

Bernreuter, D.L., Savy, J.B., Mensing, R.W., and Chung, D.H. (1984), Seismic Hazard Characterization of the Eastern United States: Methodology and Interim Results for Ten Sites, NUREG/CR-3756.

Bernreuter, D.L., Savy, J.B., Mensing, R.W., Chen, J.C., and Davis, B.C., Seismic Hazard Characterization of the Eastern United States, Volume 1: Methodology and Results for Ten Sites, UCID-20421, Vols. 1 and 2.

Chung, D.H. and Bernreuter, D.L., (1981), "Regional Relationships Among Earthquake Magnitude Scales," Reviews of Geophysics and Space Physics, Vol. 19, 649-663, see also NUREG/CR-1457 (UCRL-52745).

EPRI (1985), Seismic Hazard Methodology for Nuclear Facilities in the Eastern United States: Preliminary Seismic Hazard Test Computations for Parametric Analysis and Comparative Evaluations, EPRI Research Project Number PI01-2g (Draft).

EPRI (1986) (Electric Power Research Institute), Seismic Hazard Methodology for the Central and Eastern United States, 9 Volumes, EPRI-NP-4726.

Appendix B

Maps of the Seismic Zonation for Each of the 11 Seismicity Experts

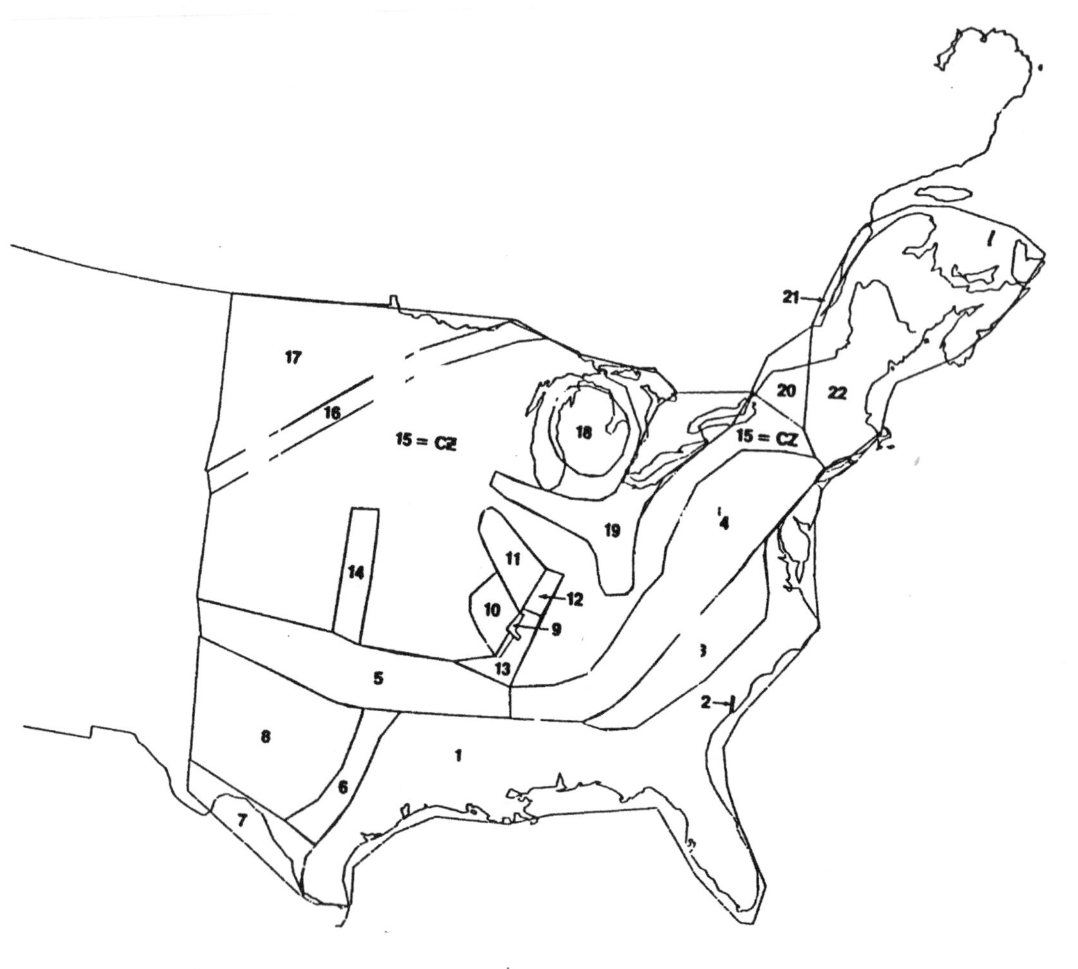

Figure B1.1 Seismic zonation base map for Expert 1.

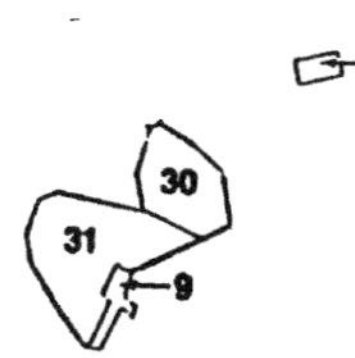

Figure B1.2 Map of alternative seismic zonation to Expert 1's base map.

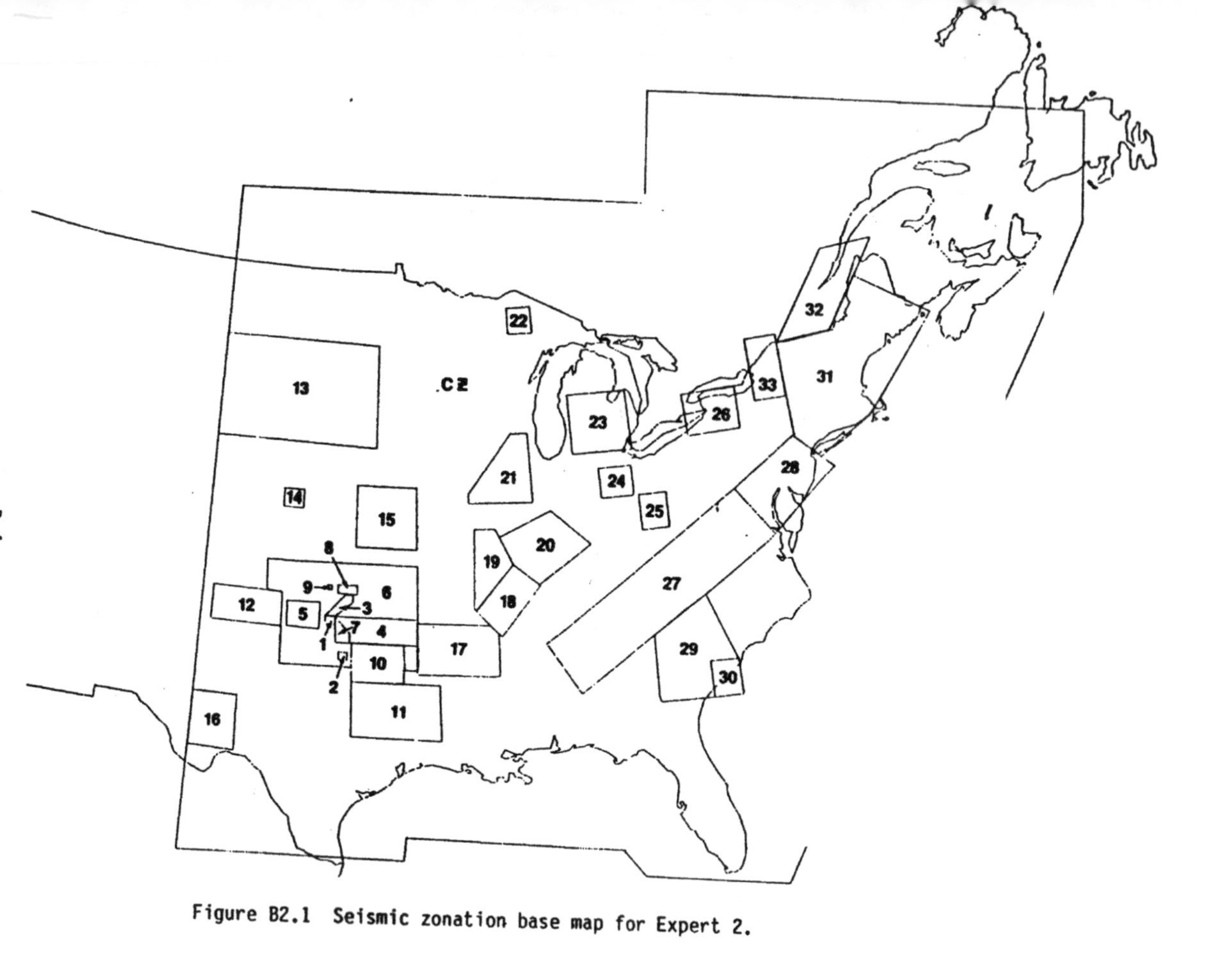

Figure B2.1 Seismic zonation base map for Expert 2.

2

16

Figure B3.1 Seismic zonation base map for Expert 3.

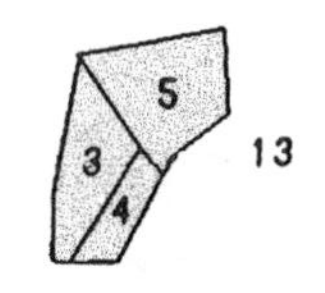

Figure B4.1 Seismic zonation base map for Expert 4.

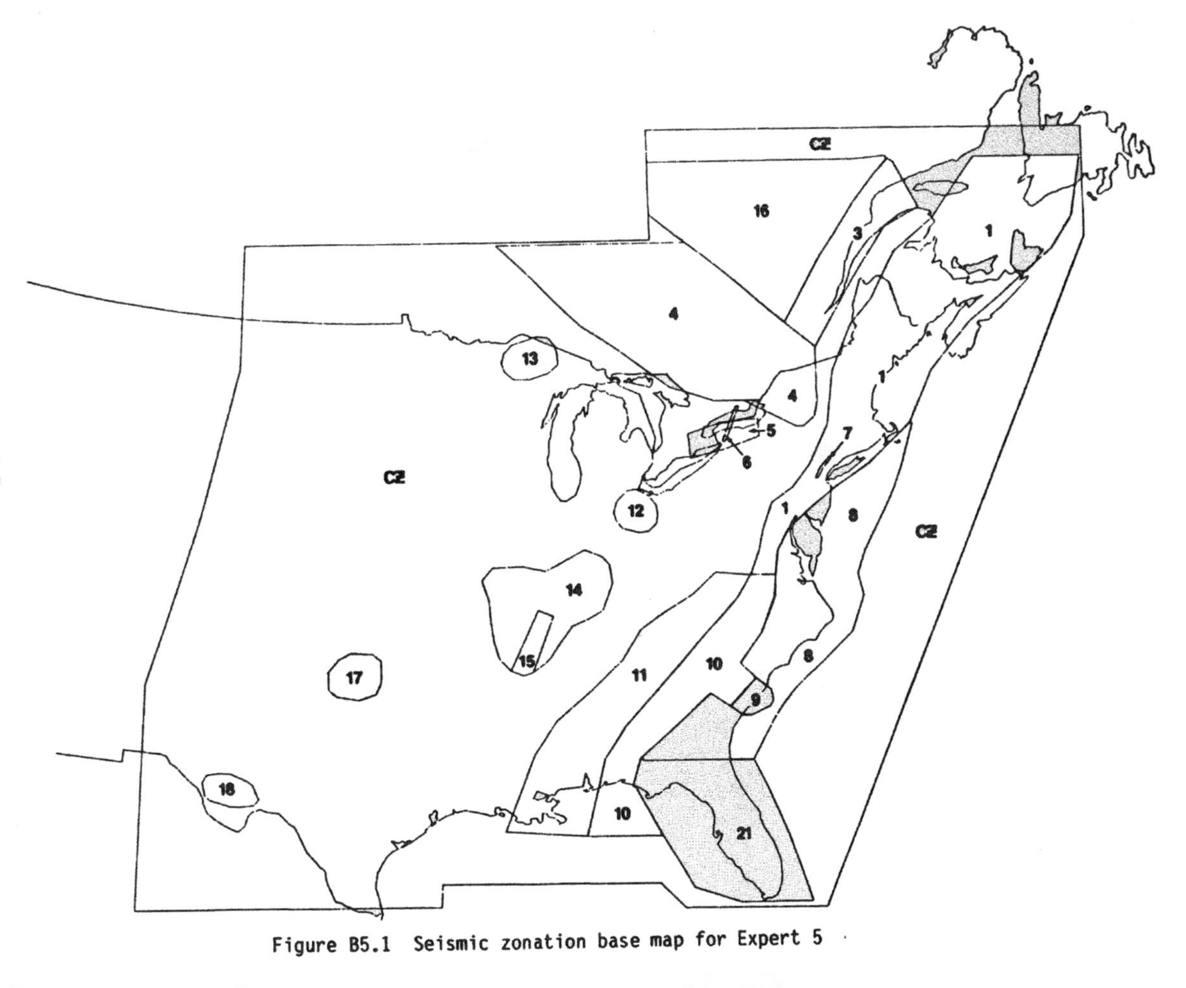

Figure B5.1 Seismic zonation base map for Expert 5

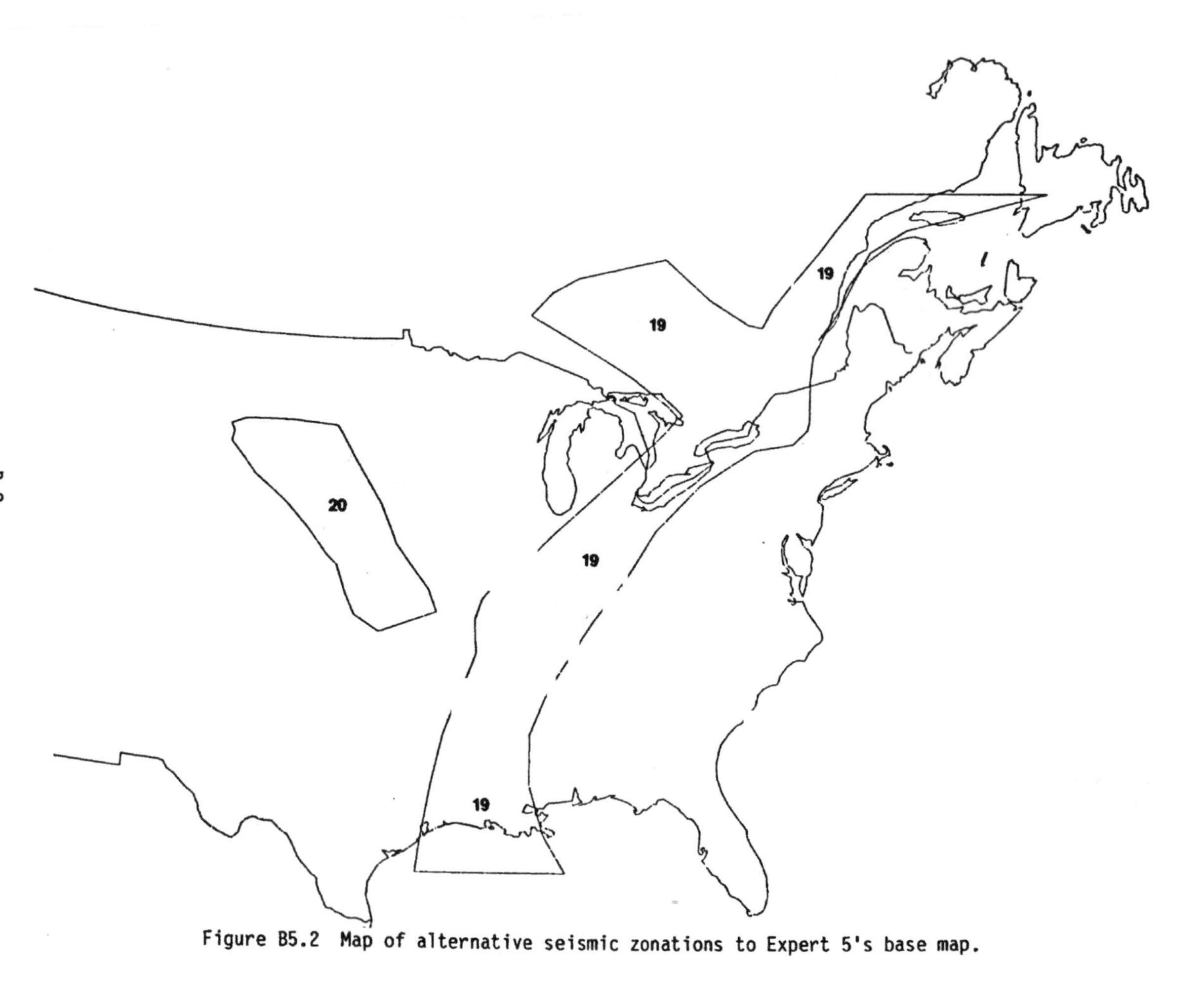

Figure B5.2 Map of alternative seismic zonations to Expert 5's base map.

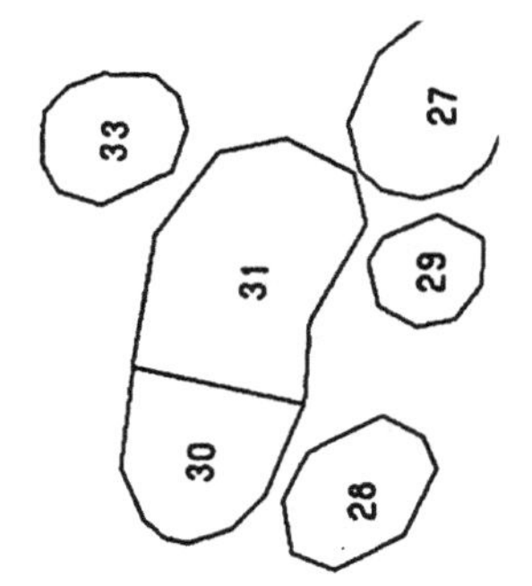

Figure B6.1 Seismic zonation

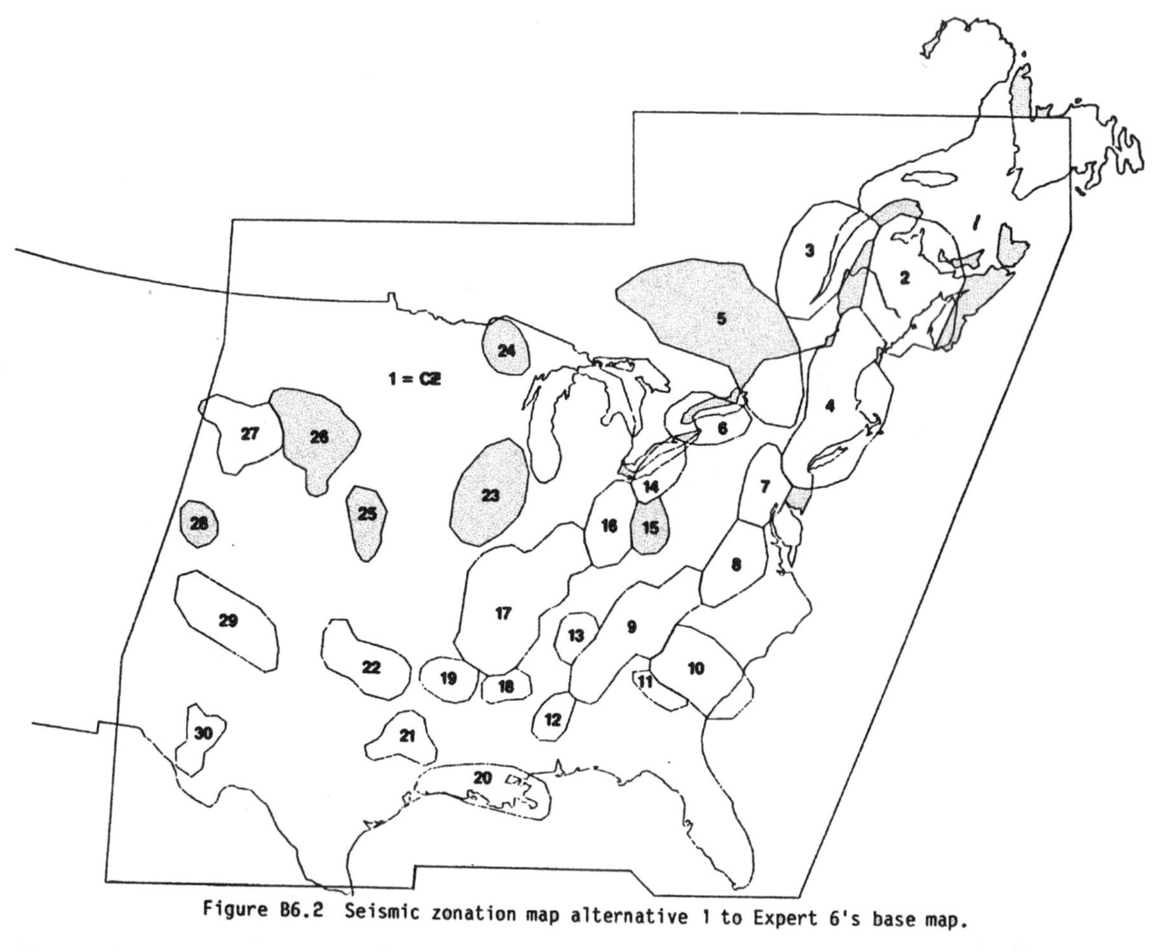

Figure B6.2 Seismic zonation map alternative 1 to Expert 6's base map.

15

14

0

Figure 7.1 Seismic zonation base map for Expert 7.

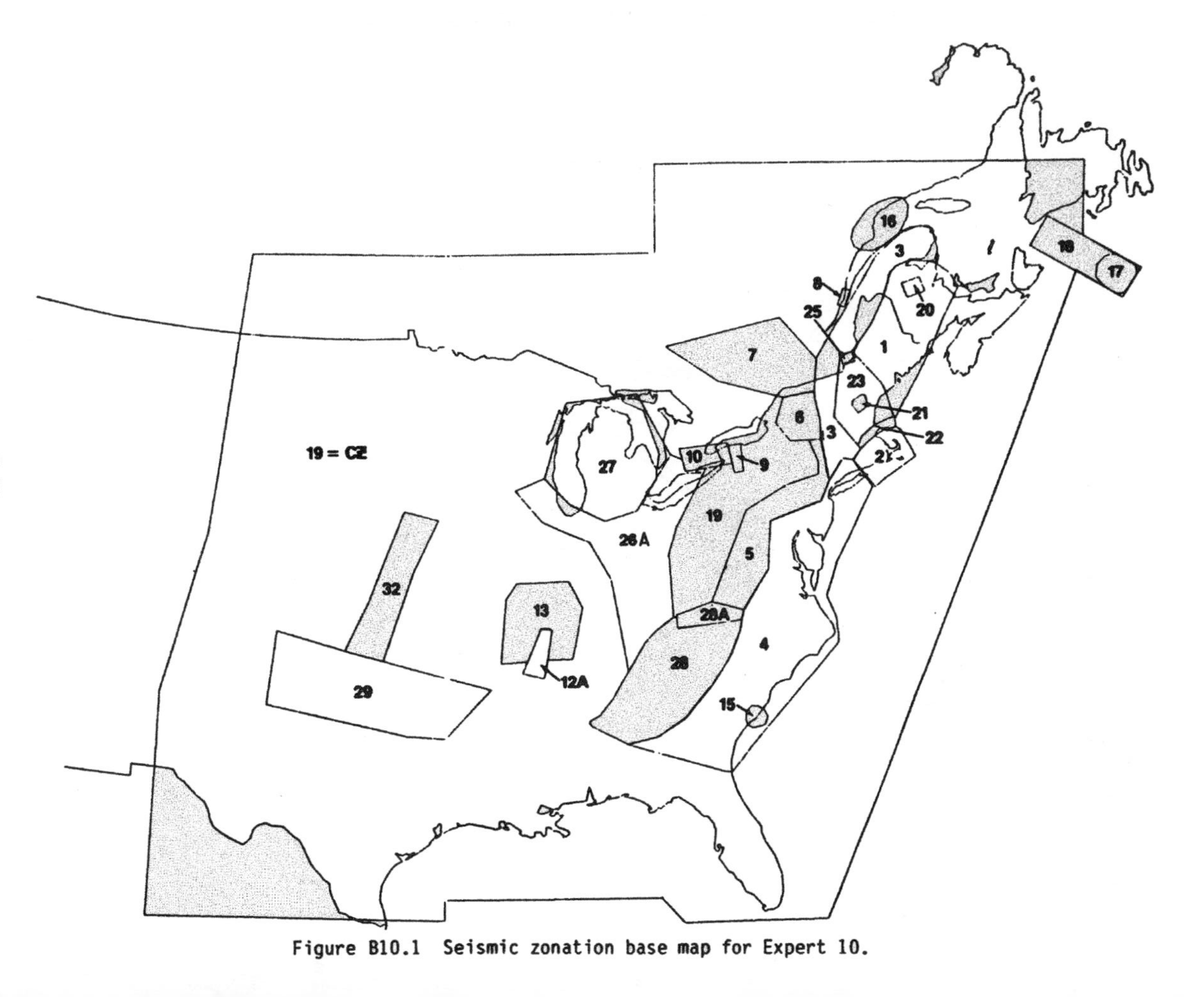

Figure B10.1 Seismic zonation base map for Expert 10.

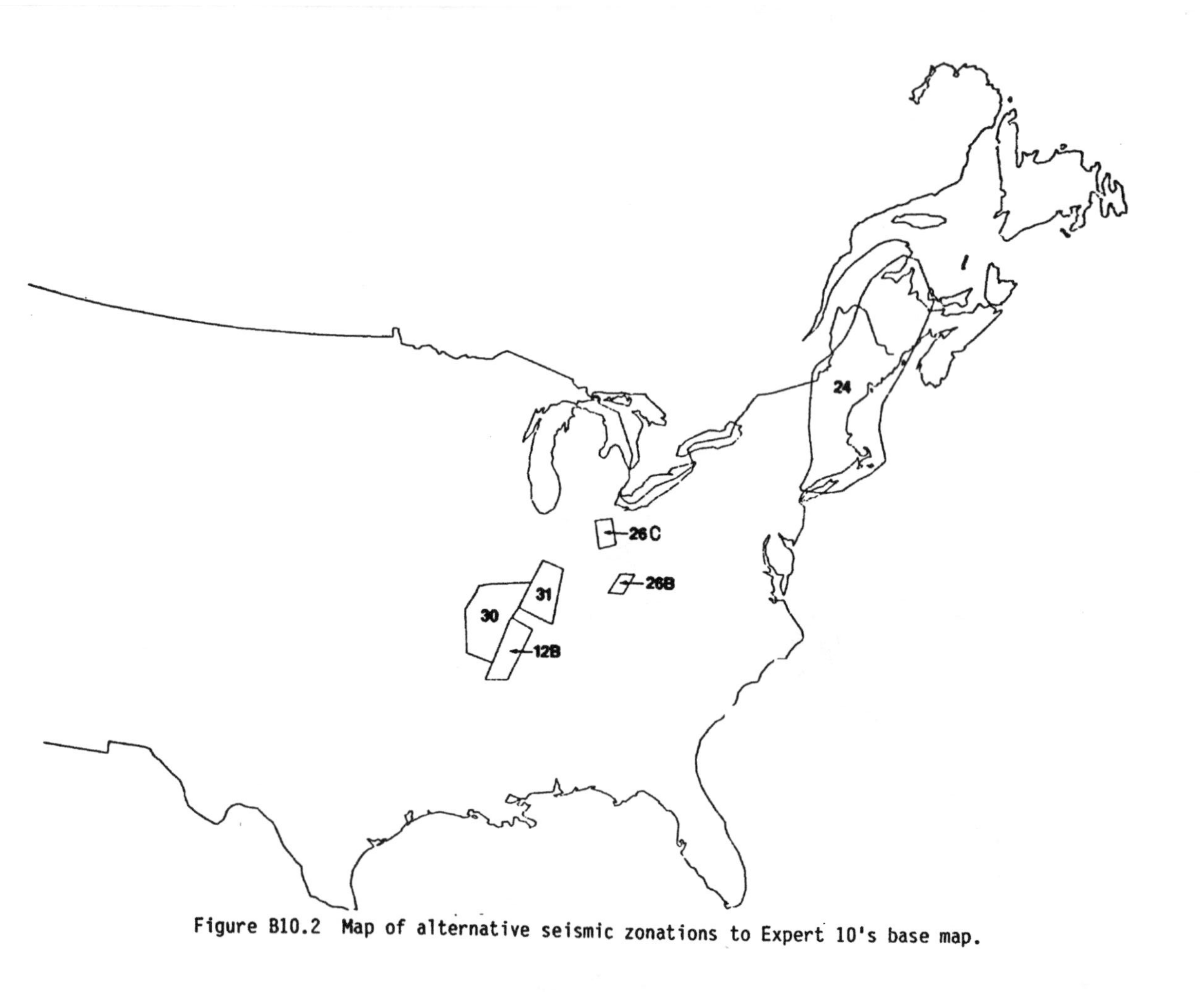

Figure B10.2 Map of alternative seismic zonations to Expert 10's base map.

0=CZ

CZ

14

Figure B11.1 Seismic zonation base map for Expert 11.

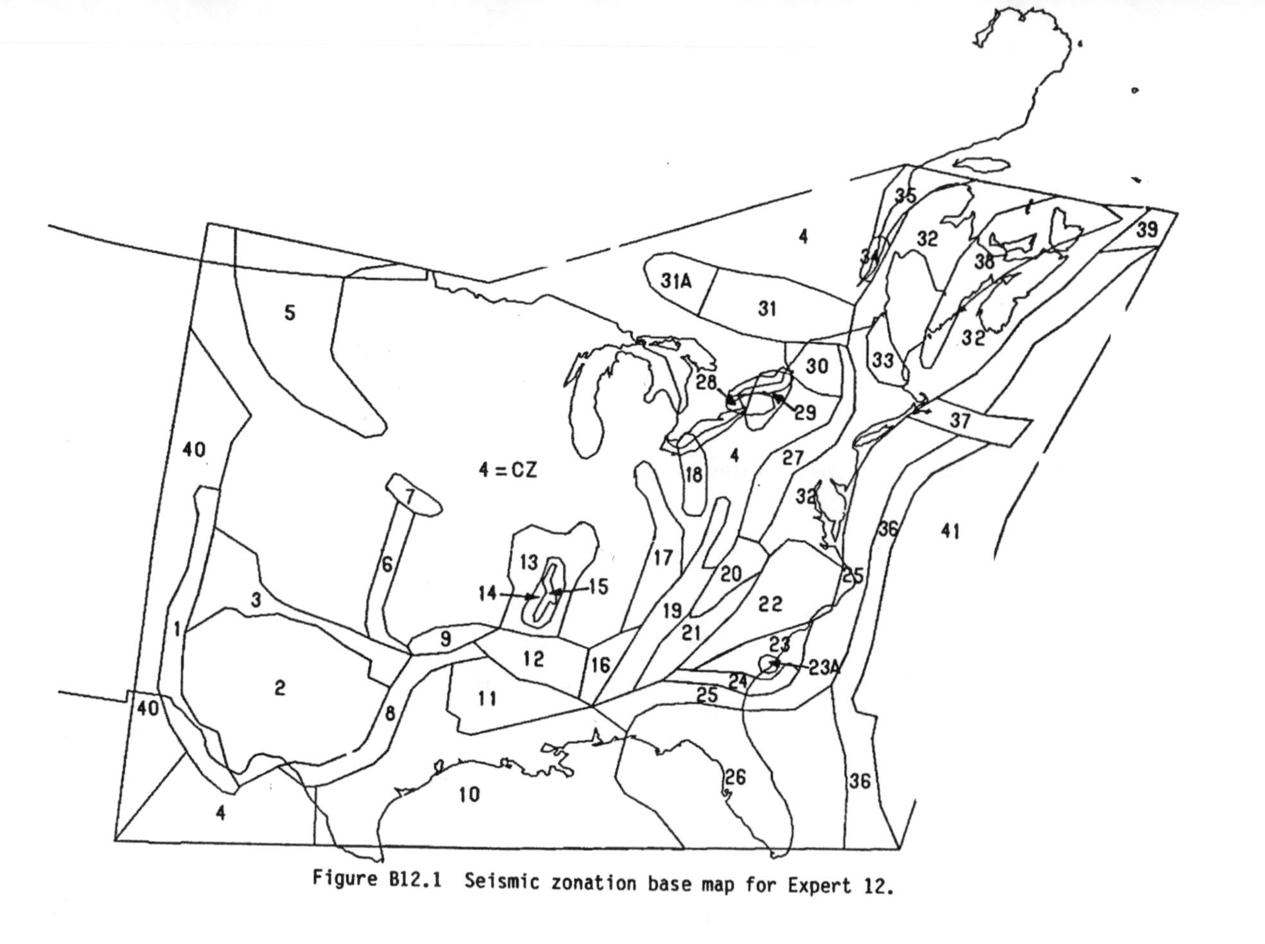

Figure B12.1 Seismic zonation base map for Expert 12.

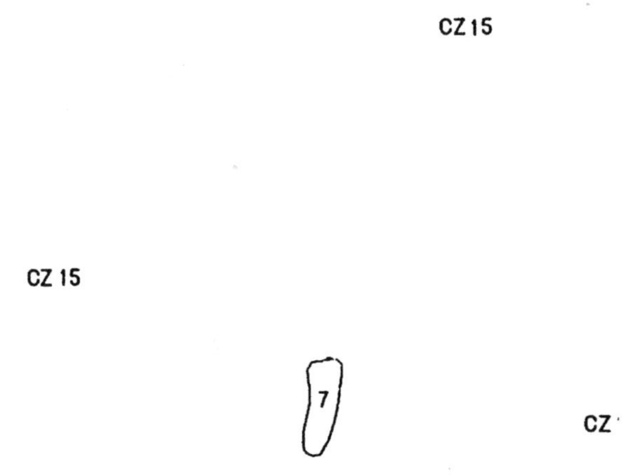

CZ15

CZ 15

7

CZ

Figure B13.1 Seismic zonation base map for Expert 13.

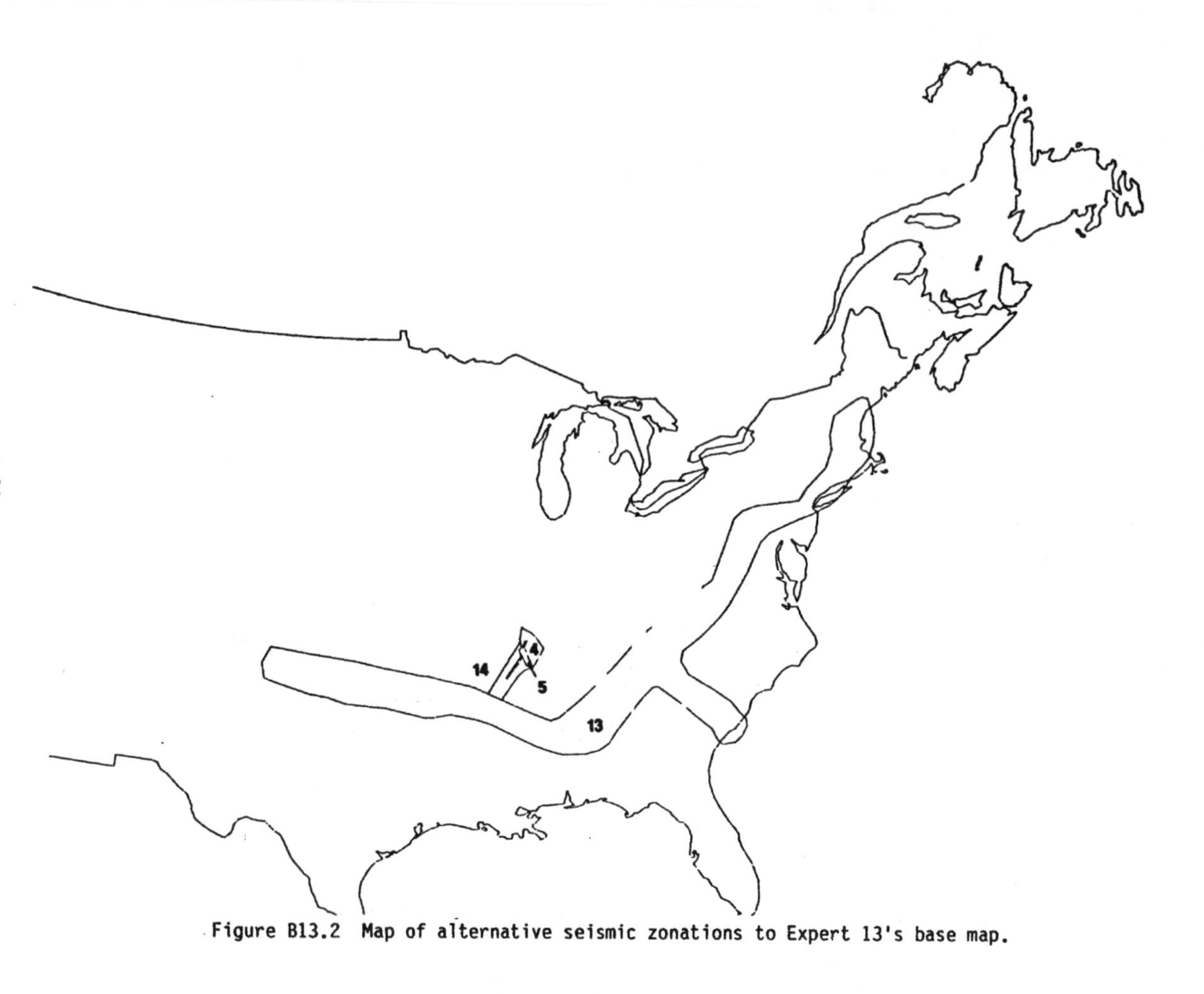

Figure B13.2 Map of alternative seismic zonations to Expert 13's base map.

7 PERFORMING ORGANIZATION NAME AND MAILING ADDRESS (Include Zip Code)

Lawrence Livermore National Laboratory
P.O. Box 808, L-197
Livermore, California 94550

10 SPONSORING ORGANIZATION NAME AND MAILING ADDRESS (Include Zip Code)

Division of Engineering and System Technology
Office of Nuclear Reactor Regulation
U.S. Nuclear Regulatory Commission
Washington, DC 20555

13 ABSTRACT (200 words or less)

The EUS Seismic Hazard Characterization Project (SHC) is the outgrowth of an earlier performed as part of the U.S. Nuclear Regulatory Commission's (NRC) Systematic Evalu Program (SEP). The objectives of the SHC were: (1) to develop a seismic hazard char ation methodology for the region east of the Rocky Mountains (EUS), and (2) the appl of the methodology to 69 site locations, some of them with several local soil condit The method developed uses expert opinions to obtain the input to the analyses. An i aspect of the elicitation of the expert opinion process was the holding of two feedb meetings with all the experts in order to finalize the methodology and the input dat bases. The hazard estimates are reported in terms of peak ground acceleration (PGA) damping velocity response spectra (PSV).

A total of eight volumes make up this report which contains a thorough description o methodology, the expert opinion's elicitation process, the input data base as well a discussion, comparison and summary volume (Volume VI).

Consistent with previous analyses, this study finds that there are large uncertainti associated with the estimates of seismic hazard in the EUS, and it identifies the gr motion modeling as the prime contributor to those uncertainties.

The data bases and software are made available to the NRC and to the public uses thr the National Energy Software Center (Argonne, Illinois).

14 DOCUMENT ANALYSIS – a. KEYWORDS/DESCRIPTORS

Seismic hazard, Eastern U.S., ground motion

b. IDENTIFIERS/OPEN ENDED TERMS

CPSIA information can be obtained
at www.ICGtesting.com
Printed in the USA
BVHW04*1530280918
528775BV00008B/219/P